# Climate Change and its Impacts: Ground Realities

# Climate Change and its Impacts: Ground Realities

**S. Jeevananda Reddy**
Hyderabad.

BSP **BS Publications**
A unit of **BSP Books Pvt. Ltd.**
4-4-309/316, Giriraj Lane, Sultan Bazar,
Hyderabad - 500 095
Phone : 040 - 23445605, 23445688

**Climate Change and its Impacts: Ground Realities**
***by Dr. S. Jeevananda Reddy***



***Published by t:***

**BSP** **BS Publications**
A unit of **BSP Books Pvt. Ltd.**

4-4-309/316, Giriraj Lane, Sultan Bazar,
Hyderabad - 500 095
Phone : 040 - 23445605, 23445688
**e-mail : info@bspbooks.net**

**ISBN: 978-93-5230-151-5 (HB)**

# Preface

Earth's climate is dynamic and it is always changing through the natural cycles. What we are experiencing now is part of this system. Also, general circulation patterns over different parts of the globe are part of this. They are highly region specific systems. However, with the unabated population growth and to meet their needs under the new lifestyles based on innovative technologies, local and regional environments were/are modified and thus they influence the natural cycles in climate at local and regional scales is represented by trend. World Meteorological Organization of United Nations brought out a manual on "Climate Change" wherein it discussed methods to separate natural inbuilt variations from man induced trend.

People of all walks of life make statements attributing weather aberrations to global warming forgetting the fact that "When was, the period, in human history in which some person, in some location or other, wasn't feeling that some recent day, month or season has been very unusual in some dry, wet, hot, or cold?" Several civilizations have gone in to the history. Most unfortunate thing is that people are shy of using the word "Global Warming" with narrow implications of emission related rise in temperature and instead use "Climate Change" as defacto global warming, a word with broader implications. Climate relates to all meteorological parameters while global warming relates to temperature only, though it is not global in nature.

The impact of unusual events of the same intensity is now felt more with population growth, infrastructure development and with intensive urbanization. Scientists started attributing weather aberration impacts on production to climate change used as de-facto global warming. Short term changes in weather and climate are different from long-term change or trend in climate that vary with space and time. With this, nowadays everything has turned into unusual and try to attribute to something they don't have the comprehensive knowledge like a blind man using a light pole. Most unfortunate thing is that man on the street to Editor of scientific magazines talk of climate change at the same wavelength, and thus science of climate change became a scapegoat. In 2016 US Presidential elections, politics are running around the climate change. The "Science of Climate Change" has turned in to political satire of "Global warming & Carbon Credits". IPCC is giving weight to number rather than to science. All these are happening even after IPCC changed its tone and tenor from AR4 to AR5 on global warming and its impact issues after 2009 December Copenhagen fiasco.

This has resulted that the nations, more particularly developing countries, wasting their meager resources on issues of little consequences to their lives, present and future. We are neglecting the pollution factor which was raised by Pope Francis in their Encyclical report. The same was also emphasized by UN and US President Obama. This is affecting the life forms on the Earth, more particularly Human Health. With this people are spending major share of their earnings towards health bills. This has been indirectly harming the environment with the drug manufacturing and associated pollution; and in its research introducing more and more new diseases and thus more research and thus more pollution and more energy use; and more pollution, the vicious circle moving on and on.

Unfortunately neither UN nor the US or Pope Francis have put realistic emphasis on pollution reduction as with this component there are severe ramifications with multinational companies [MNCs] interests. MNCs lobbied with the groups preparing the draft to be presented at Paris meet relating to COP21 in November 30 – December 12, 2015 and were successful. Because of the same reason only, UN shifted the environmental movement of 70s towards global warming and carbon emissions at 1992 Rio Summit and since then no turning back and as a result pollution component has taken back seat. Pollution has several components, namely air, water, soil & food in association with agriculture, transport, industries, urbanization, etc. Carbon dioxide is not pollution. Carbon dioxide has more positive impacts.

However, it is cautioned here that heuristic estimates based on global warming models may lead nowhere but provide a means of sensationalize inaccurate-misleading conclusions. We must realize the fact that "Ignorance is terrible but exaggeration is dangerous". After December 2009 Copenhagen "Climategate", UN Secretary General Ban Ki-Moon said "IPCC should have stronger scientific basis for making it predictions". Here the basic meteorology is not given its due place and thus lead tosensational reports. The main culprits are the media.

The theory is that there is global warming. Global warming influences ice melt thus rise in sea levels and thus influences the habitats along the coastal zones and thus affect environment. Global warming influences monsoons and thus global water cycle and thus water resources availability and thus affects agriculture, etc. The question is: Is it a fact? The book tried to answer this. Some of the materials used in the book were taken from internet sources. I herewith acknowledge those unknown authors with thanks. Also to get clarity sometimes the same material/argument is repeated.

***Dr. S. Jeevananda Reddy***

## Contents

# List of Tables and Figures

# List of Figures

1 Chapter

# General Issues in Climate Change

## 1.1 Introduction

**General:** The environmental movements started in 19$^{th}$ century and got momentum with publication of Aldo Leopold's "A Sand Country Almanac" in 1940. He believed in a land ethic that recognized that maintaining the 'beauty, integrity, and health' of natural systems" as a moral and ethical imperative; another important book in the promotion of the environmental movement was Rachel Carson's "Silent Spring" in 1962 about declining bird populations due to DDT, an insecticide, pollution and man's attempts to control nature through use of synthetic substances. Both of these books helped bring the issues into the public eye. Rachel Carson's "Silent Spring" sold over two million copies. A book by Stan Cox's "Sick Planet: Corporate Food and Medicine", in 2008 argues that the corporate food and medicine industries are destroying environments and ruining living conditions across the world.

In 1972, the United Nations Conference on the Human Environment was held in Stockholm, and for the first time united the representatives of multiple governments in discussion relating to the state of the global environment. The [Late] Indira Gandhi then the Prime Minister of India Chaired the Session. This conference has led directly to the creation of environmental agencies by individual national governments and the United Nations Environment Program [UNEP]. This has culminated introduction of environmental Acts in India also – in 1974 Water Act and lead the establishment of a environmental ministry,

Pollution Control Boards at the Centre and in the States, in 1981 Air Act & in 1986 the Environmental Act and with this presented action plans on wide range of subjects relating to Environmental issues. Unfortunately after UN Rio Summit, the environmental issue was sidelined by “one point goal”, namely “Global Warming and carbon credits” with huge money pocket to share.

Since 1970s, public awareness, environmental sciences, ecology, and technology have advanced to include modern focus points like ozone depletion, global climate change, acid rain, and the potentially harmful chemical input agriculture technology & Genetically Modified Organisms (GMOs). Now the use of IT has taken the centre stage in degrading the environment with its high energy consumption and generation of huge e-waste globally in addition creating severe human ethical & health problems.

**WMO [World Meteorological Organization of United Nations]:** To provide a framework for international cooperation in the development of meteorology and operational hydrology and their practical application in 1873 founded World Meteorological Organization [WMO] and established under UN in 1950. WMO became the specialized agency of the UN for meteorology (weather & climate), operational hydrology and related geophysical sciences. It has also the responsibility on the state and behavior of the Earth’s atmosphere, its interaction with oceans, the climate it produces and the resulting distribution of water resources. Meteorological services of UN member states are members of WMO. It produced a manual on “Climate Change” in 1966. Authors of this manual [**WMO, 1966**] are eminent meteorologists from different National Meteorological Services. [Late] Shri. K. N. Rao from India Meteorological Department [IMD] was a co-author of this manual.

**IPCC [Intergovernmental Panel on Climate Change]:** IPCC was established by WMO & UNEP in 1988/89 for the assessment of climate change. This is to provide the world with a clear scientific view on the current state of knowledge in climate change and its potential environmental and socio-economic impacts. Unfortunately, this is now serving the political perspective of climate change rather than scientific perspective of climate change. To achieve this goal, they picked up thousands of people for preparing reports. Though the title is “climate change” but in reality it is dealing primarily with “global warming and carbon credits" by spending billions of dollars each year. IPCC and others, thus, are pursuing a political agenda and PR Campaign, not scientific inquiry.

**Conference of Parties (COP):** The international political response to climate change began at the Rio Earth Summit in 1992, where the 'Rio Convention' included the adoption of the UN Framework convention on Climate Change (UNFCCC). This convention set out a framework for action aimed at stabilizing atmospheric concentrations of greenhouse gases (GHGs) to avoid "dangerous anthropogenic interference with the climate system". The UNFCCC which entered into force on 21 March 1994 now has a near-universal membership of 195 parties. As part of this game, IPCC, a political body, provides the teeth to the UNFCCC.

The main objective of the annual Conference of Parties (COP) is to review the Convention's implementation. The first COP took place in Berlin in 1995 and significant meetings since then have included COP3 where the Kyoto Protocol was adopted, COP11 where the Montreal Action Plan was produced, COP15 in Copenhagen where an agreement to success Kyoto Protocol was unfortunately not realized and COP17 in Durban where the Green Climate Fund was created. That means all roads are leading to "Fund collection & distribution".

The 2015 United Nations Climate Change Conference, COP21 was held in Paris, from November 30 to December 12. The conference was attended by about **50,000 participants including around 25,000 official delegates** from government, intergovernmental organizations, UN agencies, NGOs and civil society. According to the organizing committee, the objective of the 2015 conference is to achieve, for the first time in over 20 years of UN negotiations, a binding and universal agreement on climate, from all the nations of the world. However, the New Paris Agreement of 12$^{th}$ December 2015 has no legally binding clause. In 1992 UNFCCC laid down broad legal structure for global cooperation to which future agreements were intended to provide more specificity. Instead the Paris Agreement introduces a new, and mainly worrisome, model of voluntary "national determined contributions" by governments in terms of emissions and finances. It included a clause on limiting global temperature rise over the pre-industrial levels as "holding the increase in the global average temperature to well below 2 $^{o}$C and pursue efforts temperature increase to 1.5 °C", but this is not associated with the anthropogenic greenhouse gases but a combination of factors which are not addressed in the agreement. COP21 Agreement diluted the main object of COP21.

Paris Agreement Document observed that "Recognizing the intrinsic relationship between climate change, poverty eradication and equitable access to sustainable development, and reaffirming responses to climate change should aim to meet the specific needs and concerns arising from the adverse impacts of response measures". They used all types of jargons: "creation of decent work and quality jobs development priorities; safeguarding food security and ending hunger, and the particular vulnerabilities of food production systems to the adverse impacts of climate change; promoting, protecting and respecting all human rights, the right to health, and the rights of indigenous peoples, migrants, children, persons with disabilities and people in vulnerable situations and under occupation, and the right to development, in accordance with their obligations, as well as promoting gender equality and the empowerment of women, when taking action to address climate change, Noting the needs and integrity of terrestrial ecosystems, oceans and Mother Earth". Though all these refer to "climate change and pollution aspects", in reality they were not addressed in the final document. The document dealt only on global warming and emission control.

Pope Francis published an encyclical called 'Laudato Si' intended, in part, to influence the conference. The encyclical calls for action against "human-caused climate change". The International Trade Union Confederation has called for the goal to be "zero carbon, zero poverty", and the general secretary Sharan Burrow has repeated that there are "no jobs on a dead planet". This conference itself has produced substantial amount of greenhouse gases by spending huge sums for travel and stay.

In this process, Philippe Verdier, Weather Chief at France Televisions, the country's state broadcaster, has been suspended for publicly criticizing climate alarmism. Mr. Verdier claims in the book Climate Investigation that leading climatologists and political leaders have "taken the world hostage" with misleading data. He added: "We are hostage to planetary scandal over climate change – a war machine whose aim is to keep us in fear". This is the present state of atmosphere on climate change issues.

**Greenpeace:** Greenpeace [a pro-global warming group] Founder [now left the organization after serving 15 years] delivered a lecture on 14th October 2015 in London. He observed that "As I have stated publicly on many occasions, there is no definitive scientific proof, through real-world observation, that carbon dioxide is responsible for any of the slight warming of the global climate that has occurred during the past 300 years, since the peak of the Little Ice Age. If there were such a proof through testing and replication it would have been written down for all to see. The contention that human emissions are now the dominant influence on climate is simply a hypothesis, rather than a universally accepted scientific theory. It is therefore correct, indeed verging on compulsory in the scientific tradition, to be skeptical of those who express certainty that 'the science is settled' and 'the debate is over'. But there is certainty beyond any doubt that $CO_2$ is the building block for all life on the Earth and that without its presence in the global atmosphere at a sufficient concentration this would be a dead planet."

When scientists questioned them on these pro-global warming groups dubbed such scientists as agents of fuel companies, instead answering their point of views. The present book's objective is to discuss all issues relating to climate, to present the issue of climate change in right perspective.

## 1.2 General Issues of Climate Change

### 1.2.1 Weather and Climate

**Weather** is basically the way the atmosphere is behaving, mainly with respect to its effects upon life and human activities. Weather is not the same everywhere. Perhaps it is hot, dry and sunny today where you live, but in other parts of the world it is cloudy, raining or even snowing. Every day, weather events are recorded and predicted by meteorologists worldwide. In order to help people be prepared to face all of these, National Weather Service [NWS], in India we have "Indian Meteorological Department" [IMD], are the lead forecasting outlet for the nation's weather. They also provide Special Weather Statements and Short and long term Forecasts. NWS also issues a lot of notices concerning marine weather for boaters and others who dwell or are staying near shoreline.

Weather is the mix of events that happen each day in our atmosphere that include meteorological parameters such as temperature, humidity, precipitation, cloudiness, brightness, visibility, wind, and atmospheric pressure. There are really a lot of other components to weather, namely sunshine, radiation, evaporation, rain, cloud cover, winds, hail, snow, sleet, freezing rain, flooding, blizzards, ice storms, thunderstorms, steady rains from a cold front or warm front, excess heat, heat waves, excess cold, cold waves and more.

**Climate** is the average weather pattern at a place. In most places, weather can change from minute to minute, hour-to-hour, day-to-day, season-to-season and year-to-year. Climate, however, is the average of such weather over time and or space. An easy way to remember the difference is that "climate is what you expect, like a very hot summer, and weather is what you get, like a hot day with pop-up Thunderstorms". Meteorologists record the weather every day at specific times or continuously using automatic recorders. Climate data is useful for weather forecasting and is useful to define location or region in terms of climate condition. The Earth's climate is influenced by many factors, including solar radiation, wind, and ocean currents. But interaction among the various factors is very complex and numerous questions remain unresolved.

Weather research is concerned with the formation, movement, and prediction of the individual elements of weather, such as a particular low-pressure system and a hurricane. Climate research, on the other hand, deals with the more comprehensive totality of low pressure systems and hurricanes, and is dedicated to addressing questions such as how many mid-latitudinal storms or hurricanes will occur next year, or whether they will become more frequent or intense in the coming years. So the term "weather" refers to short-term events in the atmosphere, while "climate" relates to longer time periods.

Climate, sometimes understood as the "average weather," is defined as the measurement of the mean and variability of relevant quantities of certain variables (such as temperature, precipitation or wind) over a period of time, ranging from months to thousands or millions of years. The classical period is 30 years, as defined by the World Meteorological Organization (WMO). Climate is the characteristic condition of the atmosphere near the Earth's surface at a certain place on the Earth. Two of the most important factors determining an area's climate are air temperature and precipitation. Every NMS published Climate normal books that contain averages and extremes for individual meteorological stations. IMD published normal book using 1931 to 1960 met data – this is known as "Red Book".

**Table 1.1** presents the averages and extremes for widely distributed 46 Indian meteorological stations for temperature and precipitation along with their latitudes, longitude and elevation [above mean sea level (amsl)]. For the highest and the lowest rainfall, data even before 1931 was also taken in to account. Rainfall includes the annual average in mm, number of average rainy days, the highest and the lowest averages along with the year of occurrence and the maximum amount of rainfall received in 24 hours; and in the case of temperature, presented the highest maximum, the mean maximum, the mean minimum and the lowest minimum. They are highly variable with the space and the time.

**Table 1.1** Climate averages and extremes for 46 locations in India [a] Station location, [b] Precipitation and [c] temperature

| **[a] Station location** | | | |
|---|---|---|---|
| **Station** | **Latitude ° ' N** | **Longitude ° ' E** | **Elevation [amsl, m]** |
| Dibrugarh | 27 28 | 94 55 | 106 m |
| Gauhati | 26 11 | 9145 | 55 |
| Shillong | 25 34 | 91 53 | 1500 |
| Cherrapunji | 28 15 | 91 44 | 1313 |
| Darjeeling | 27 03 | 88 16 | 2127 |
| Jalpaiguri | 26 32 | 88 43 | 83 |
| Assansol | 23 41 | 86 59 | 126 |
| Calcutta | 22 32 | 88 20 | 6 |
| Sambalpur | 21 28 | 83 58 | 148 |
| Gopalpur | 19 16 | 84 53 | 17 |
| Daltanganj | 24 03 | 84 04 | 221 |
| Jamshadpur | 22 49 | 86 11 | 129 |
| Patna | 25 37 | 85 10 | 53 |
| Lucknow | 26 52 | 80 56 | 111 |
| Fatehpur | 25 56 | 80 50 | 114 |
| Mussoorie | 30 27 | 78 05 | 2042 |

*Contd...*

| Station | Latitude ° ' N | Longitude ° ' E | Elevation [amsl, m] |
|---|---|---|---|
| Mukteswar | 29 28 | 79 39 | 2311 |
| Dehra Dun | 30 19 | 78 02 | 682 |
| Roorkee | 29 51 | 77 53 | 274 |
| Ludhiana | 30 56 | 75 52 | 247 |
| Hissar | 29 10 | 75 44 | 221 |
| New Delhi | 28 35 | 77 12 | 216 |
| Leh | 34 09 | 77 34 | 3514 |
| Srinagar | 34 05 | 74 50 | 1586 |
| Jammu | 32 40 | 74 50 | 366 |
| Barmer | 25 45 | 71 23 | 194 |
| Gwalior | 26 14 | 76 15 | 207 |
| Indore | 22 43 | 75 48 | 567 |
| Dwaraka | 22 21 | 69 05 | 11 |
| Alibag | 18 38 | 72 52 | 7 |
| Poona | 1832 | 73 51 | 559 |
| Mahabaleswar | 17 56 | 73 40 | 1382 |
| Sholapur | 17 40 | 75 54 | 479 |
| Hyderabad | 17 27 | 78 28 | 545 |
| Anantapur | 14 41 | 77 37 | 350 |
| Visakhapatnam | 17 43 | 83 14 | 3 |
| Madras | 13 00 | 80 11 | 16 |
| Coonoor | 11 21 | 76 48 | 1747 |
| Coimbatore | 11 00 | 76 58 | 409 |
| Nagapattinam | 10 46 | 79 51 | 9 |
| Tiruchirapalli | 10 46 | 78 43 | 88 |
| Kodaikanal | 10 14 | 77 28 | 2343 |
| Pamban | 09 16 | 79 18 | 11 |
| Mangalore | 12 52 | 74 51 | 22 |
| Bellary | 15 09 | 76 51 | 419 |
| Trivandrum | 09 29 | 76 57 | 64 |

*Contd...*

| [b] Precipitation | | | | | |
|---|---|---|---|---|---|
| Station hours | Annual Mean [mm] | Rainy days | The Highest [mm]/year | The Lowest [mm]/year | Max. in 24 hours |
| Dibrugarh | 2759.4 | 131.7 | 3300.2/1954 | 2165.0/1905 | 223.5 |
| Gauhati | 1637.2 | 86.9 | 2121.1/1956 | 874.0/1952 | 232.9 |
| Shillong | 2415.3 | 128.1 | 3334.0/1946 | 1428.8/1958 | 415.5 |
| Cherrapunji | 11418.7 | 157.1 | 15706.6/1951 | 178.0/1908 | 973.0 |
| Darjeeling | 2758.4 | 125.1 | 4024.4/1890 | 2271.3/1907 | 492.8 |
| Jalpaiguri | 3352.7 | 103.7 | 4292.2/1938 | 1719.6/1891 | 390.4 |
| Assansol | 1392.2 | 72.5 | 2135.1/1922 | 963.7/1955 | 269.7 |
| Calcutta | 1581.8 | 83.7 | 2501.4/1893 | 909.1/1935 | 369.1 |
| Sambalpur | 1661.5 | 75.4 | 2307.8/1929 | 0934.7/1915 | 401.3 |
| Gopalpur | 1209.5 | 55.7 | 2040.9/1936 | 0655.1/1920 | 510.8 |
| Daltonganj | 1234.4 | 67.4 | 1856.5/1946 | 0622.3/1903 | 290.8 |
| Jamshedpur | 1391.3 | 75.7 | 1862.7/1953 | 0745.5/1954 | 214.1 |
| Patna | 1109.8 | 53.8 | 1959.4/1918 | 0642.4/1903 | 366.0 |
| Lucknow | 0992.4 | 47.9 | 1866.7/1915 | 0424.2/1907 | 311.7 |
| Fatehpur | 0885.1 | 48.1 | 1360.7/1936 | 0537.2/1941 | 191.8 |
| Mussoorie | 2368.1 | 94.6 | 3284.7/1942 | 1475.7/1929 | 302.3 |
| Mukteswar | 1359.4 | 80.6 | 2254.2/1936 | 0522.9/1951 | 254.5 |
| Dehra Dun | 2313.7 | 85.4 | 3118.9/1884 | 1152.4/1907 | 332.2 |
| Roorkee | 1164.6 | 51.2 | 2299.7/1942 | 0511.6/1935 | 266.7 |
| Ludhiana | 0704.5 | 37.0 | 1402.1/1955 | 0236.5/1899 | 254.3 |
| Hissar | 0446.0 | 27.2 | 1048.0/1917 | 0158.0/1938 | 346.7 |
| New Delhi | 0714.2 | 36.8 | 1533.1/1933 | 0262.9/1929 | 266.2 |
| Leh | 0115.0 | 13.5 | 0231.1/1894 | 0025.4/1881 | 005.3 |
| Srinagar | 0664.0 | 56.9 | 1291.6/1894 | 0398.5/1934 | 147.8 |
| Jammu | 1148.3 | 52.3 | 1964.7/1948 | 0646.4/1940 | 276.1 |
| Barmer | 0310.3 | 17.1 | 0895.3/1944 | 0128.8/1938 | 255.5 |
| Gwalior | 0900.2 | 44.0 | 1514.3/1934 | 0224.0/1918 | 316.2 |
| Indore | 1053.4 | 50.3 | 1743.2/1959 | 0400.1/1899 | 293.4 |
| Dwaraka | 0418.9 | 17.0 | 1080.0/1944 | 0026.2/1918 | 355.1 |
| Alibag | 2124.9 | 82.2 | 3396.7/1958 | 0720.6/1941 | 396.2 |
| Poona | 0714.7 | 50.1 | 1242.3/1892 | 0268.5/1918 | 149.1 |
| Mahabaleswar | 6182.3 | 122.7 | 10221.2/1896 | 3545.1/1899 | 339.9 |
| Sholapur | 0742.0 | 46.5 | 1239.5/1916 | 0325.4/1899 | 191.0 |
| Hyderabad | 0764.4 | 52.3 | 1430.8/1915 | 0455.2/1899 | 190.5 |
| Anantapur | 0562.3 | 34.0 | 0777.9/1919 | 0233.2/1934 | 145.3 |

*Contd...*

| Station hours | Annual Mean [mm] | Rainy days | The Highest [mm]/year | The Lowest [mm]/year | Max. in 24 hours |
|---|---|---|---|---|---|
| Visakhapatnam | 0973.6 | 50.1 | 1442.0/1910 | 0473.2/1925 | 293.3 |
| Madras | 1215.3 | 56.6 | 2134.9/1943 | 0522.2/1904 | 261.6 |
| Coonoor | 1549.2 | 84.3 | 2239.0/1960 | 1130.3/1950 | 228.6 |
| Coimbatore | 0612.2 | 44.5 | 1059.2/1924 | 0323.6/1938 | 141.5 |
| Nagapattinam | 1336.5 | 53.5 | 2196.3/1884 | 0603.5/1897 | 396.2 |
| Tiruchirapalli | 0867.6 | 47.8 | 1324.1/1939 | 0511.6/1914 | 319.0 |
| Kodaikanal | 1672.1 | 107.0 | 2354.6/1925 | 1184.1/1904 | 346.2 |
| Pamban | 0922.9 | 45.2 | 1722.2/1896 | 0410.5/1892 | 218.7 |
| Mangalore | 3467.0 | 122.7 | 4703.1/1946 | 2269.7/1899 | 360.9 |
| Bellary | 0518.1 | 33.6 | 0949.4/1933 | 0208.3/1884 | 162.3 |
| Trivandrum | 1839.3 | 105.2 | 3035.6/1933 | 1029.2/1894 | 277.9 |

**[c] Air Temperature [°C]**

| Station | The Highest | Maximum | Minimum | The Lowest |
|---|---|---|---|---|
| Dibrugarh | 36.5 | 27.7 | 18.7 | 6.9 |
| Gauhati | 37.0 | 29.5 | 19.7 | 7.8 |
| Shillong | 28.2 | 21.2 | 12.1 | -0.6 |
| Cherrapunji | 27.5 | 20.6 | 14.3 | 4.0 |
| Darjeeling | 23.5 | 16.4 | 10.2 | -0.9 |
| Jalpaiguri | 36.3 | 28.8 | 19.3 | 7.2 |
| Assansol | 45.1 | 32.0 | 20.8 | 8.0 |
| Calcutta | 41.1 | 31.8 | 22.1 | 9.6 |
| Sambalpur | 45.4 | 32.9 | 20.8 | 7.2 |
| Gopalpur | 37.0 | 30.3 | 23.0 | 12.7 |
| Daltonganj | 45.1 | 31.8 | 19.0 | 3.9 |
| Jamshedpur | 44.8 | 32.4 | 20.7 | 6.9 |
| Patna | 43.7 | 31.6 | 20.8 | 6.5 |
| Lucknow | 45.5 | 32.3 | 19.4 | 4.3 |
| Fatehpur | 45.9 | 32.4 | 19.5 | 3.8 |
| Mussoorie | 29.8 | 18.0 | 10.4 | - 2.7 |
| Mukteswar | 27.8 | 17.6 | 09.5 | - 3.5 |
| Dehra Dun | 40.9 | 27.9 | 15.8 | 2.1 |
| Roorkee | 44.1 | 30.2 | 17.0 | 2.2 |
| Ludhiana | 46.4 | 31.9 | 17.3 | 1.4 |
| Hissar | 46.3 | 32.8 | 17.4 | 0.8 |

*Contd...*

| Station | The Highest | Maximum | Minimum | The Lowest |
|---|---|---|---|---|
| New Delhi | 44.9 | 31.7 | 18.8 | 2.7 |
| Leh | 30.0 | 12.4 | - 1.4 | - 20.8 |
| Srinagar | 36.0 | 19.5 | 07.2 | - 7.7 |
| Jammu | 45.2 | 30.0 | 18.7 | 4.4 |
| Barmer | 46.3 | 34.0 | 22.2 | 4.9 |
| Gwalior | 46.4 | 32.5 | 18.8 | 1.3 |
| Indore | 42.5 | 31.3 | 17.5 | 3.3 |
| Dwaraka | 37.1 | 29.2 | 22.9 | 11.6 |
| Alibag | 36.8 | 30.0 | 22.6 | 13.4 |
| Poona | 41.6 | 32.0 | 18.2 | 6.2 |
| Mahabaleswar | 33.4 | 24.1 | 16.1 | 8.4 |
| Sholapur | 43.3 | 33.7 | 20.5 | 10.0 |
| Hyderabad | 42.4 | 31.7 | 20.0 | 9.1 |
| Anantapur | 41.4 | 33.3 | 21.9 | 13.2 |
| Visakhapatnam | 39.6 | 31.0 | 23.5 | 15.2 |
| Madras | 41.5 | 32.9 | 24.3 | 17.3 |
| Coonoor | 27.0 | 21.2 | 12.6 | 3.7 |
| Coimbatore | 37.5 | 31.1 | 21.7 | 15.8 |
| Nagapattinam | 40.0 | 31.9 | 25.2 | 19.3 |
| Tiruchirapalli | 40.5 | 33.7 | 24.0 | 17.2 |
| Kodaikanal | 22.9 | 17.9 | 10.8 | 4.7 |
| Pamban | 34.1 | 30.6 | 25.7 | 21.5 |
| Mangalore | 34.9 | 30.5 | 23.7 | 18.6 |
| Bellary | 41.1 | 32.9 | 22.3 | 14.0 |
| Trivandrum | 34.9 | 30.7 | 23.5 | 19.6 |

### 1.2.2 Climate System-Climate Change Scenario

**Climate System**: The climate system consists of five major components [**Figure 1.1**], namely the atmosphere, the hydrosphere, the cryosphere, land surface, the biosphere. The climate system is continually changing due to the interactions between the components as well as external factors such as volcanic eruptions or solar variations and human-induced factors such as changes to the atmosphere and changes in land use. The atmosphere is not an isolated system. It interacts with other components of the Earth system – the oceans, for example. But it is also in contact with the cryosphere (ice and snow), the biosphere (animals and plants), the pedosphere (soil) and the lithosphere (rocks). All of these elements together compose the climate system, whose individual components and processes are connected and influence each other in diverse ways.

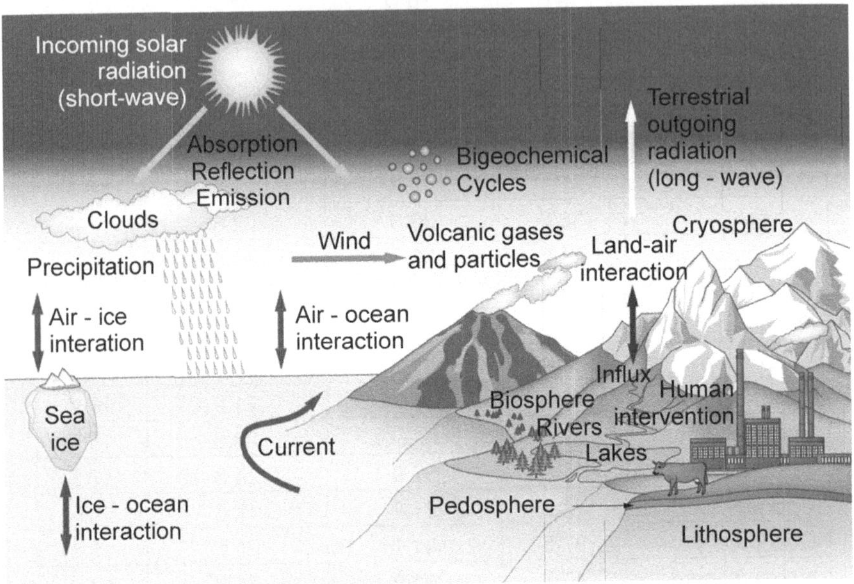


**Figure 1.1** Climate System

A large continental ice mass such as the Antarctic ice sheet, as a result of climate change, presumably undergoes change over many millennia, and without counteractive measures it will gradually melt on this time scale. The predictability of climate is based on the interactions between the atmosphere and the more inert climate subsystems, particularly the oceans. Within this scheme, the various components of the climate system move at completely different rates. Low-pressure systems can drift hundreds of kilometers within days. Ocean currents, on the other hand, often creep along at a few meters per minute. In addition, the individual components possess different thermal conductivities and heat capacities. Water, for instance, stores large amounts of solar heat for long periods of time.

**Climate Variability:** Climate Variability is defined as variations in the mean state and other statistics of the climate on all temporal and spatial scales, beyond individual weather events. The term "Climate Variability" is often used to denote deviations of climatic statistics over a given period of time (e.g. a month, season or year) when compared to long-term statistics for the same calendar period. Climate variability is measured by these deviations, which are usually termed anomalies. Variability may be due to natural internal processes within the climate system (internal variability), or to variations in natural or anthropogenic external factors (external variability).

**Climate Change:** According to United Nation entities climate change has not one, but several definitions. In fact they are nothing new from what has presented in **WMO [1966]** "climate change" manual but made some twists to create confusion. Let us see two such definitions:

1. The IPCC Third Assessment Report defines climate change: "Climate change refers to a statistically significant variation in either the mean state of the climate or in its variability, persisting for an extended period [typically decades or longer]. Climate change may be due to natural internal processes or external forcings or to persistent anthropogenic changes in the composition of the atmosphere or in land use". According to the IPCC, climate change can occur naturally or from man-made causes.
2. That IPCC definition, however, goes on to read: Note that the UN Framework Convention on Climate Change [UNFCCC], in its Article 1, defines "climate change" as: "a change of climate which is attributed *directly or indirectly* to human activity that alters the composition of the *global atmosphere* and which is in addition to natural climate variability observed over comparable time periods".

The UNFCCC thus makes a distinction between "climate change" attributable to human activities altering the atmospheric composition, and "climate variability" attributable to natural causes. Under external forcings: In the first definition IPCC differentiated between anthropogenic changes in the composition of the atmosphere from land use. In the second IPCC used human activity that alters the composition of the global atmosphere.

In the first definition they used simply "atmosphere" and in the second it used "global atmosphere"; In the first they referred to "land use" and this was differentiated from anthropogenic [meaning new additions of greenhouse gas]; in the second they used "human activity" but before it, they used "directly or indirectly" may be to differentiate greenhouse gases from land use, to create confusion. All these are clearly defined by WMO in 1966. IPCC to meet its global warming and carbon credit policy, it goes on creating confusion from one report to the other.

**Global warming** --- Earth Observatory/NASA presented an essay on Global Warming. It state that "Throughout its long history, the Earth has warmed and cooled time and again. Climate has changed when the planet received more or less sunlight due to subtle shifts in its orbit, as the atmosphere or surface changed, or when the Sun's energy varied. But in the past century, another force has started to influence the Earth's climate, namely humanity. Global warming is the unusually rapid increase in the Earth's average surface temperature over the past century "primarily due to greenhouse gases" related as people burn fossil fuels.

How does this warming compare to previous changes in the Earth's climate? How can we be certain that human-released greenhouse gases are causing the warming? How much more will the Earth warm? How will the Earth respond? Answering these questions is perhaps the most significant scientific challenge of our time.

### 1.2.3 Is Global Warming a Reality?

The main component of climate change is natural variability. Humans have no control over it and thus we need to adapt to them **[Reddy, 1993]**. They are location-region specific like that of "general circulation patterns". Extremes are part of natural variability. This can be seen from the climate normal data of temperature & precipitation [**Table 1.1**] wherein the current extremes have not crossed those past recorded limits up to now. Also, regional general circulation patterns over different seasons play the key role in year to year variations in extremes under natural variability. For example Western Disturbances in northwestern parts of India will influence heat and cold waves in summer and winter based on the High Pressure belt location around Nagpur– east and west and north and south shift with the passing of time in any given year -- [**Figure 1.2 --** Figure 7 a & b in **Reddy & Rao, 1978**]. The winds associated with the high pressure belt define the penetration of cold winds in winter and warm winds in summer in to southern parts and eastern parts of India.

WEATHER ASSOCIATED WITH WESTERN DISTURBANCES

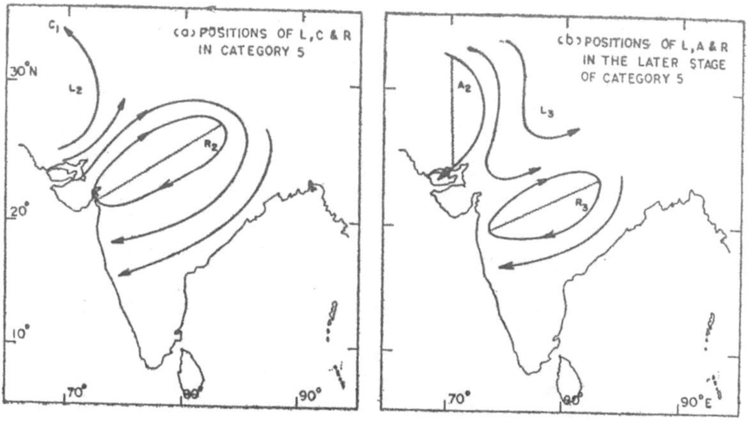


**Note:** Where L denotes the surface low associated with Western disturbance, A or C denotes the anti-cyclonic or cyclonic flow at 0.9 km and R the ridge at 850 mb level. A or C does not refer to the centre of the system but are used only to indicate anti-cyclonic or cyclonic flow

**Figure 1.2** Position of L, A, C and R in category 5 [weather associated with western disturbance]

The second important component of climate change is "ecological changes" – associated with changes in land use & land cover and water use & water cover changes. They are highly location-region specific and are visible with naked eye. "**Urban-Heat-Island Effect**" and "**Rural-Cold-Island Effect**" comes under this. They are expressed in terms of trend [increasing or decreasing] while natural variability is expressed by cyclic variation or rhythmic variation or by near sine curve, called fluctuations. The trend will influence the natural rhythmic variation present in that location-region. This plays an important role in power consumption in urban areas and agriculture in rural areas.

The third is "Greenhouse Effect" related to temperature. This is a natural change. Atmosphere consists of several gases. Through hydrological cycle [**Figure 1.3**], the water vapour component changes with season and latitude, precisely with general circulation patterns and climate system. The Sun emits radiation in different wavelength bands [**Figure 1.4**]. Some of these gases in the atmosphere interact with radiation in different wavelength bands. Similarly, the Earth re-radiates energy in long wave radiation. This also interacts with the atmospheric gases. The wavelength at which the maximum energy is concentrated depends upon the energy emitting body's temperature. The Sun radiates energy at 6000 $^{o}$K and the Earth at 10 $^{o}$K. According to Plank's Law, the Sun's energy is concentrated at around 0.5μ [short wave] and the Earth's energy at around 10μ [long wave]. This process of interaction, we call it as greenhouse effect. This converts the energy in to temperature. Due to man's actions certain gases like $CO_2$ are added to the atmosphere. The associated temperature raise is termed as "Global Warming".

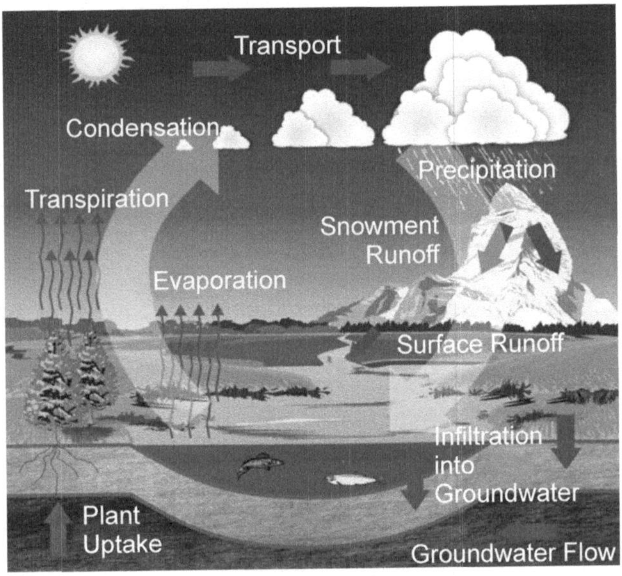


**Figure 1.3** Hydrological cycle

The fourth one is associated with the atmospheric aerosols pumped in to atmosphere [troposphere and stratosphere] by several natural [volcanic eruptions, Earthquakes] and manmade [wars, dust storms, nuclear explosions, etc] activities. They both have warming and cooling effects similar to heat-island and cold-island effects. They are localized/regional factors may have short or long life depending upon the intensity of such activity. Because of their highly intermittent in nature, it makes complications in assessing their affect in true perspective. Here we must remember one important point: while cooling, the energy available to greenhouse effect is reduced and thus warming is proportionately reduced as warming is a function of greenhouse gases present in the atmosphere and energy available at that time. Thus, the contribution to warming will be reduced.

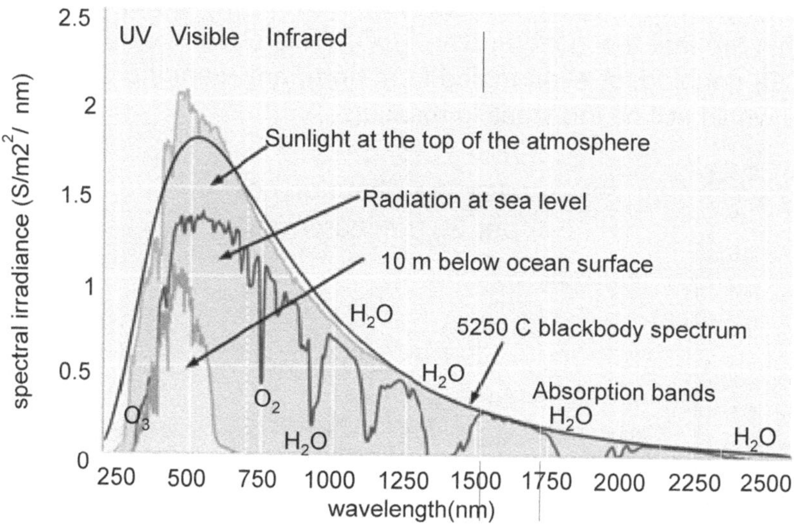


**Figure 1.4** Sun's Energy spectrum under different wavelength bands

IPCC and others argued that global warming is a settled science but we feel it is not so basically because: in nature the conversion component of natural $CO_2$ in to temperature has reached a near saturation **[Reddy, 1995a]**. Any new addition to $CO_2$ due to man's actions, the increase in temperature will be insignificant. **Figure 1.5** presents the Heating Effect of $CO_2$ per 20 ppm increments.

Because of this, IPCC with 97% scientific support [as they claim] is using "trial and error" approach to link $CO_2$ raise to temperature raise. As a part of this game, IPCC goes on changing the sensitivity factor that relates $CO_2$ with temperature – in AR4 used 1.95 and in AR5 it reduces 1.95 to 1.55 – SAR to AR5 the sensitivity factor showed a monotonic decline. Unfortunately over this mean the range show very wide as ± 50% around the mean. In science it has no meaning. Even the model predictions show plateau pattern even before 2100. Even it is there this will follow a tapering pattern. That is the rise in temperature gradually decreases with time for the same amount of increase in $CO_2$. Dbstealey [28 January 2016] noted that "But so far, no one has produced any measurements quantifying AGW. That means either: (1) global warming does not exist, or (2) global warming is too minuscule to measure. It is down in the noise. I think global warming exists [**Figure 1.5**], but that most of the effect took place within the first few dozen ppm of $CO_2$. This chart shows that even if $CO_2$ doubled, or even tripled from here, any warming due to that rise would still be too small to measure.

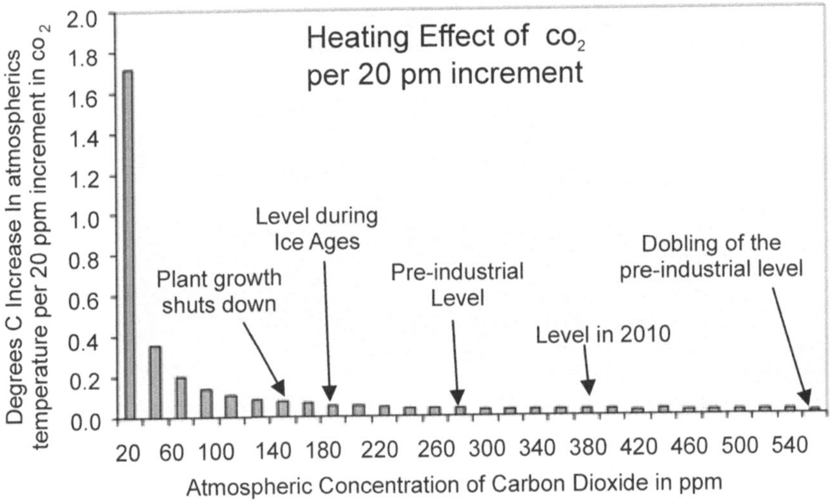


**Figure 1.5** Heating effect of $CO_2$ per 20 ppm increments.

Some argued that at $CO_2$ absorption wavelengths less energy is escaping to space. They cited satellite data to prove this. However, at the same time the satellite temperature data presented small increase in global temperature anomaly since 1979 to date. This does not support the reports that "enhanced greenhouse effect" but supports opposite to this – tapering-off with the time as it is seen in **Figure 1.5**.

Also, IPCC now states that global warming plus other forcings is more than half of the global temperature rise [50.1% is also more than half] -- but COP21 declaration uses the global average temperature rise over pre-industrial level [IPCC used 1951 as the starting year for global warming]. This is a subjective statement. IPCC has no valid data to verify its estimates as the global temperature curve is built on hypothetically derived data series over around 75-80% of the area through interpolations or extrapolations under grids.

When the current data series showed a hiatus for the last 19 years, three different theories were put forward – one group manipulates the data to remove hiatus, another group says temperature is hiding in deep-ocean and the third says it is a part of natural variation [Figure 1.6a]. The data refers to ground based measurements. Bob Tisdale observed that "NOAA revised their global surface temperature product in June 2015 to show more global warming during the post-1998 period. Those data manipulations supposedly ended slowdown in global warming over that period. The changes in NOAA's global ocean surface temperature product were the primarily cause of the NOAA's hiatus-disappearing Act. The new NOAA ERSST.v4 sea surface temperature data set has since been included in both the NOAA and the GISS global land + ocean surface temperature products. Because the oceans cover about 70% of the Earth's surface, the new NOAA ERSST.v4 data are critical component." More details on the data issues are presented in **Chapter 2. Figure 1.6b** presents the temperature trend of Japan.

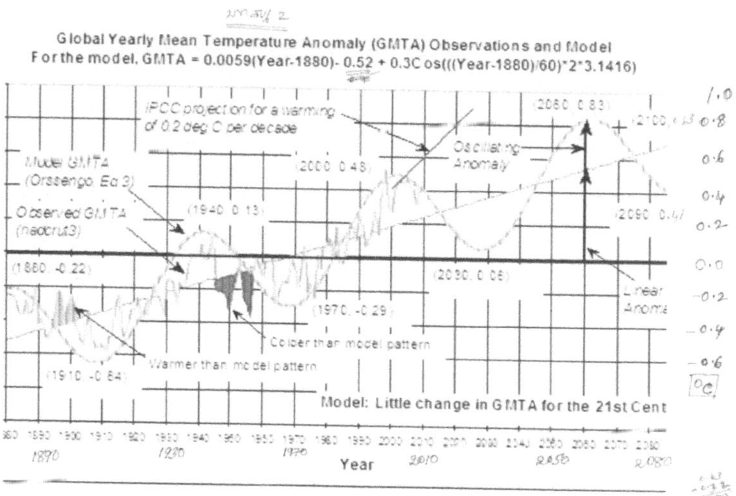


[a]

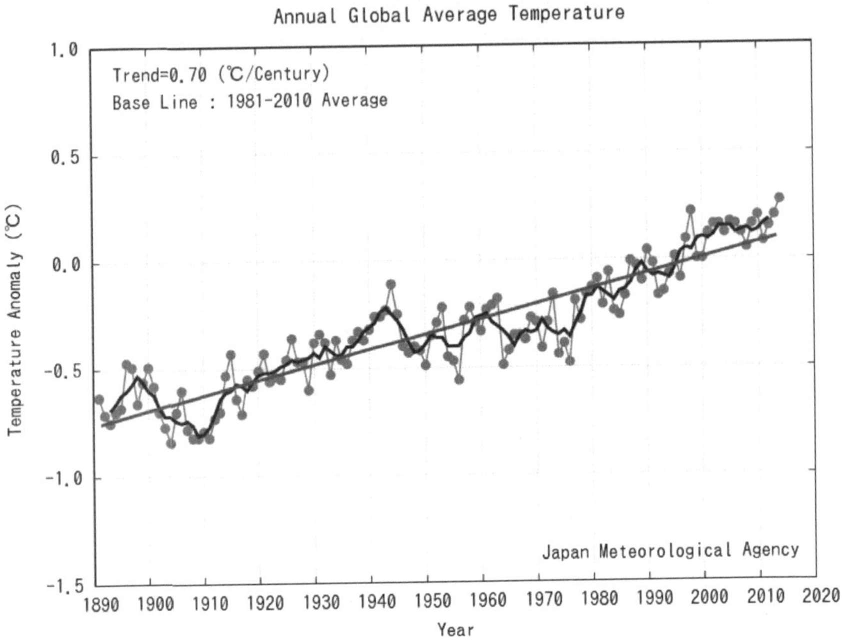


Anomalies are deviation from baseline (1981-2010 Average).
The black thin line indicates surface temperature anomaly of each year.
The blue line indicates their 5-year running mean.
The red line indicates the long-term linear trend.

**[b]**

**Figure 1.6** [a] Global temperature trend and 60-year cycle and IPCC prediction, [b] Annual global average temperature [Japan]

Let us see the temperature patterns [three surface data sets plus two satellite data sets] between 1979 and 2015. They present two distinct sets, namely prior to 1997/98 and after 1997/98 [El Nino]. “Prior part” showed a zero trend irrespective of data set and more or less all the data sets coincided. During this period volcanic eruptions were reported. As per IPCC they create cooling effect. Also the energy reaching the Earth is reduced and thus greenhouse effect component is also reduced. Thus, it presented below average condition. “After part” showed zero trends in satellite data sets but steep rise in surface data sets. Also, the five data sets patterns differed from one another. During 2014 and 2015 the anomalies respectively are ranging 0.2 to 0.7 and 0.3 to

0.9 °C [met data is shown only to the 1st place of decimal -- rounded]. The satellite data sets: 0.2 to 0.3 and 0.3 to 0.4; and surface data sets: 0.5 to 0.7 and 0.6 to 0.9 °C. In this part, however, three El Nino events influenced the temperature. With every El Nino, on either side of the peak present dip in temperature below the average condition. There is no chance that temperature anomaly presents a steep rise, which is clearly seen in satellite data series. The average from 1979 to 2015 presents in between these two zero trends – normal condition, showing a hiatus.

**Figure 1.7** presents the observed and modeled estimates of global average temperature anomaly for the period 1979-2014. In this period there were about seventeen thousands of Radiosonde Balloon temperature records in 4 data sets. They closely match the two satellite data sets. The ground based data sets derived by GISS, NOAA, and USHCN don't match, nor do climate models. According to this data series, using IPCC inference, global warming component will be around 0.15 °C from 1951 to 2015, which is insignificant to influence climate or nature. Politicians don't understand these intrinsic issues and politicians and scientists only look at 100 billon dollars – Paris summit!!! US President, Pope Francis, and others used IPCC documents as gospels.

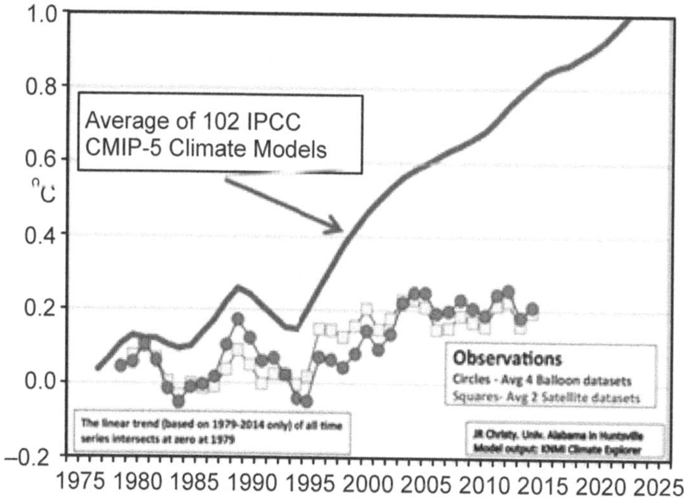


**Figure 1.7** Observed and model estimates of global average temperature anomaly

Let us see the statement --- The IEA report has worrying conclusion that the world won't be able to limit warming to 2 °C even if all the pledges (INDCs) submitted in advance of Paris Summit get implemented, "The (global) emissions trajectory implies a long-term temperature increase of 2.7 °C by 2100. A major course correction is still required to achieve the world's agreed climate goal. As the largest source of global greenhouse-gas emissions, the energy sector must be at the heart of global action to tackle climate change." --- I responded as follows:

The global [land & ocean] temperature data of 1880 to 2010 following the WMO 1966 methodology separated trend from cyclic variation [**Figure 1.6a**]. From the trend line it is clear that by 2100, the global [land & water] temperature may raise to 1.30 $^{o}$C from 1880. That is in 220 years the rise is 1.30 $^{o}$C; this is around 0.6 $^{o}$C per century, which consists of three components: One of them is emission component. Even if we take this as 50% [my suggestion was 25% only but with the time this will further reduced], then the emission component [global warming] is 0.3 $^{o}$C per century. To reach 2 $^{o}$C, the time taken is around six and half centuries [around 650 years]. In fact this global [land & water] temperature is over estimate. This is seen from satellite and balloon data [**Figure 1.7**]. According to balloon and satellite data to reach 2 $^{o}$C it will take more than 20 centuries at the present rate of emissions. We must note the fact that the groups are talking 2 $^{o}$C with reference to emission reduction only and thus 2 $^{o}$C is not the global [land & ocean] temperature rise but it should refer to global warming. Also, model estimates show a reduction with the time and reaching a plateau. Thus, the 6.5 centuries or 20 century will be doubled or tripled or even more as even in model estimates after 2100 the conversion is plateau as seen in **Figure 1.5**.

All the light emitted by the surface in the strongest $CO_2$ bands was completely absorbed at pre-industrial levels. That is by 280 ppm. Same is also evident in **Figure 1.5.** The calculations showed by reducing 30% of carbon dioxide the temperature reduction will stand at less than 0.01 to 0.02 $^{o}$C [it was around 0.003 $^{o}$C for USA alone], which is obvious from **Figures 1.5**.

Our leaders say something, talk something and act something else. There is no coherence in their speeches and to real goals on which they talk of. For example they quote sufferings of one 'X' and build the story. Finally they are falling in to the trap of global warming and $CO_2$. How this is going to save that 'X' from hunger-starvation is a big question. In the past activists brought out an album on Save Africa collected billions and really the amount went in to that component is not even 10% and the rest went in to the pockets of the activists. Same is the objective at Paris meet.

IT groups wanted to spread its roots in all aspects of life. This is highly power consuming sector of activity. Instead of putting a cap on such activities, power production cannot be reduced and thus $CO_2$. Indian PM at Silicon Valley did the same bargain and IT top group say they will dump their IT into rural India, which in turn is going to affect severely agriculture sector and health of education system. During previous PM time they wanted to spend few thousand crores on providing free cell phones to rural labour. I got it stopped on the same basis. Agriculture is the backbone to mitigate the goals UN announced and that is what US President saying and Pope Francis saying. The incoherent actions will lead nowhere except to collect 100 billion dollars and use that money to protect global warming!!!

Dr. Judith Curry's verbal testimony before the House of Representatives Committee on Science, Space and Technology hearing on the President's UN Climate Pledge observed that: "The central issue in the scientific debate on climate change is the extent to which the recent (and future) warming is caused by human-caused greenhouse gas emissions versus natural climate variability that are caused by variations from the Sun, Volcanic eruptions, and large-scale ocean circulations. Recent data and research supports the importance of natural climate variability and calls into question the conclusion that humans are the dominant cause of recent climate change. This includes [a] The slowdown in global warming since 1998; [b] Reduced estimates of the sensitivity of climate to carbon dioxide; and [c] Climate models that are predicting much more warming than has been observed so far in the 21st century.

While there are substantial uncertainties in our understanding of climate change, it is clear that humans are influencing climate in the direction of warming. However this simple truth is essentially meaningless in itself in terms of alarm, and does not mandate a particular policy response. We have made some questionable choices in defining the problem of climate change and its solution: [a] The definition of 'dangerous' climate change is ambiguous, and hypothesized catastrophic tipping points are regarded as very or extremely unlikely in the 21st century; [b] Efforts to link dangerous impacts of extreme weather events to human-caused warming are misleading and unsupported by evidence; [c] Climate change is a 'wicked problem' and ill-suited to a 'command and control' solution; and [d] It has been estimated that the U.S. national commitments to the UN to reduce emissions by 28% will prevent three hundredths of a degree centigrade in warming by 2100.

What Judith Curry said on data series that "The greatest changes in the new NOAA surface temperature analysis are to the ocean temperatures since 1998? This seems rather ironic, since this is the period where there is the greatest coverage of data with the highest quality of measurements – ARGO buoys and satellites don't show a warming trend. Nevertheless, the NOAA team finds a substantial increase in the ocean surface temperature anomaly trend since 1998."

On Australian Graham Lloyd article, "Climate scientists defend data changes – see **Figures 1.8**". The Bureau of Meteorology is quoted as claiming that "statistical analysis of minimum temperatures at Rutherglen indicated jumps in the data in 1966 and 1974..... These changes were determined through comparison with 17 nearby sites". **Figure 1.9** presents an example showing data manipulation. This at last allows me to understand how they went about turning a cooling trend of -0.33 °C per 100 years into a warming trend of +1.74 °C. I checked the monthly unadjusted minimum data for Rutherglen, the adjusted data for Rutherglen, and the unadjusted data at all 17 of the listed neighbours, in the period 1951 – 1980, which according to the Bureau is the critical period containing the 1966 and 1974 break points. 30 years is a suitably long period for analysis. For the technically minded, I calculated monthly anomalies from the 1951-1980 means for each record, and then 12 month averages. This should allow us to see the problems around 1966 and 1974. The idea of "homogenizing" was to adjust the data so that it is not so different from the neighbours. That happened in 1966. They got that right, but not in 1974, where the adjustments have *increased* the difference, and have produced warming. Odd things also happen in 1952, 1954, 1957, 1969, and 1975-80. It is clear that the changes to the temperatures at Rutherglen do not "homogenize" them. They make the differences from the neighbours greater, and change a cooling trend into a warming one. This is not unique to Rutherglen- adjustments warm the temperature trends at 66 of the 104 Australian sites, and warm the national mean temperature trend by around 47%. But what would I know- I'm just an amateur according to Professor Karoly. The raw data series for USA temperature from 1920 showed a 60-year cyclic variation. The adjusted data showed both the trend and 60 year cycle. Here the trend was introduced through manipulation of data.

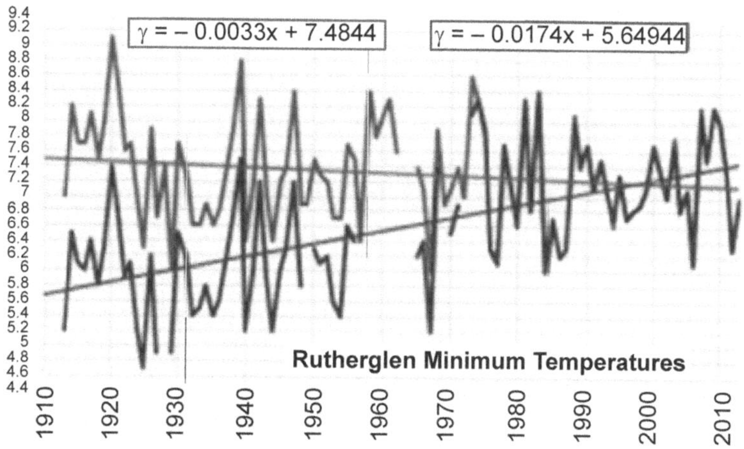


**Figure 1.8** Rutherglen temperature data [1913-2013]: unadjusted vs adjusted

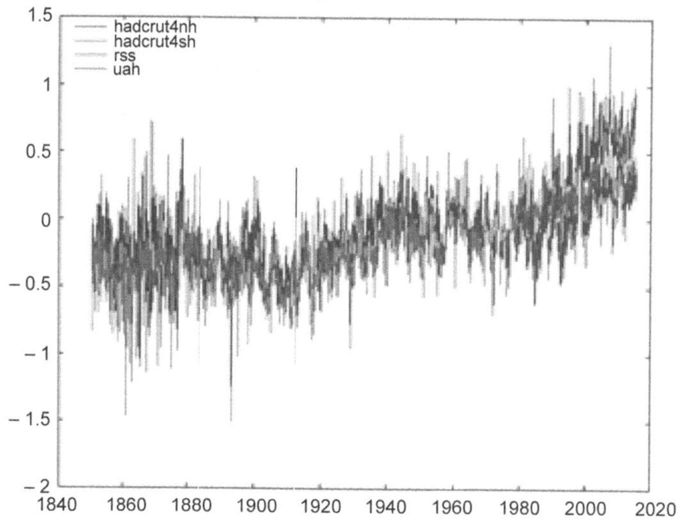


**Figure 1.9** Showing the data manipulation

According to IPCC, the global warming started in 1951. The data presented in **Table 1.1** relates to the climate conditions before 1960. The data showed wide variations with time and space. Extremes in rainfall and temperature are very high. The local life – human, animal & plant – are adapted to those wider variations in temperature and rainfall. The agriculture system was adapted to those defined local climate conditions. With the chemical input mono crop agriculture system, the problems started as they are not defined clearly to the local conditions. The loses are not much prior to 1960 as at that time the population of these factors very low but now they increased several fold and thus losses for the same extreme conditions are very high.

## 1.3 Factors Determining Earth's Temperature

### 1.3.1 General

The climate system is a complex, interactive system consisting of the atmosphere, land surface including topography/topography and vegetation, snow and ice, oceans and other bodies of water, and living things. The atmospheric component of the climate system most obviously characterizes climate; climate is often defined as 'average weather'. Climate is usually described in terms of the mean and variability of temperature, precipitation and wind over a period of time, ranging from months to millions of years (the classical period is 30 years). The climate system evolves in time under the influence of its own internal dynamics and due to changes in external factors that affect climate (called 'forcings'). External forcings include natural phenomena such as volcanic eruptions and solar variations, as well as human-induced changes.

There are feedback mechanisms in the climate system that can either amplify ('positive feedback') or diminish ('negative feedback') the effects of a change in climate forcing. For example, as rising concentrations of greenhouse gases warm Earth's climate, snow and ice begin to melt. This melting reveals darker land and water surfaces that were beneath the snow and ice, and these darker surfaces absorb more of the Sun's heat, causing more warming, which causes more melting, and so on, in a self-reinforcing cycle. This feedback loop, known as the 'ice-albedo feedback', amplifies the initial warming caused by rising levels of greenhouse gases. Detecting, understanding and accurately quantifying climate feedbacks have been the focus of a great deal of research by scientists unraveling the complexities of Earth's climate. This is hypothetical as it is not established the fact that the rising concentrations of greenhouse gases really warm Earth's climate, if so to what extent. The source of energy is the Sun. This is fixed. Greenhouse gases are there in the atmosphere. Through greenhouse effect, the Earth's temperature reached to a stable condition. Over this intra-seasonal and intra-annual, decadal changes are superposed in association with the general circulation patterns through which the climate is controlled over different parts of the globe associated with the climate system. Also, trend contains a part of long-term variations due to changes in extra-terrestrial phenomenon.

Feedbacks are an important factor in determining the sensitivity of the climate system to increased atmospheric greenhouse gas concentrations. Other factors being equal, higher climate sensitivity means that more warming will occur for a given increase in greenhouse gas forcing. Uncertainty over the effect of feedbacks is a major reason why different climate models project different magnitudes of warming for a given forcing scenario. More research is needed to understand the role of clouds and carbon cycle feedbacks in climate projections.

Weather and thus climate are represented by several meteorological parameters. Among these the principal parameters that are of importance are precipitation and temperature. They are influenced by several terrestrial and extra-terrestrial phenomena. They present high spatial and temporal variations. See **Table 1.2** for the averages and extremes in temperature and precipitation at six met stations in India. Temperature data series present high intra-seasonal and intra-annual variations. Also, it contains an anomaly around the mean. This anomaly includes both cyclic variation and trend part [**Figure 1.6a**]. The cyclic part is natural in built system and the trend part is contributed by human actions. Though this is small compared to mean, yet we need to understand this, to quantify its impacts on nature. In this Part it is tried to understand the factors contributing to the anomaly in general and trend in specific.

**Table 1. 2** Temperature and precipitation means and extremes at six locations in India: As an example

| Station | The Highest [°C] | Maximum [°C] | Minimum [°C] | The Lowest [°C] |
|---|---|---|---|---|
| Dibrugarh | 36.5 | 27.7 | 18.7 | 6.9 |
| Gauhati | 37.0 | 29.5 | 19.7 | 7.8 |
| Shillong | 28.2 | 21.2 | 12.1 | -0.6 |
| Cherrapunji | 27.5 | 20.6 | 14.3 | 4.0 |
| Darjeeling | 23.5 | 16.4 | 10.2 | -0.9 |
| Jalpaiguri | 36.3 | 28.8 | 19.3 | 7.2 |

| Station | Annual Mean [mm] | Rainy Days | The Highest [mm]/year | The Lowest [mm]/year | Max. in 24 hours |
|---|---|---|---|---|---|
| Dibrugarh | 2759.4 | 131.7 | 3300.2/1954 | 2165.0/1905 | 223.5 |
| Gauhati | 1637.2 | 86.9 | 2121.1/1956 | 874.0/1952 | 232.9 |
| Shillong | 2415.3 | 128.1 | 3334.0/1946 | 1428.8/1958 | 415.5 |
| Cherrapunji | 11418.7 | 157.1 | 15706.6/1951 | 178.0/1908 | 973.0 |
| Darjeeling | 2758.4 | 125.1 | 4024.4/1890 | 2271.3/1907 | 492.8 |
| Jalpaiguri | 3352.7 | 103.7 | 4292.2/1938 | 1719.6/1891 | 390.4 |

### 1.3.2 Temperature Data Quality

The validity of any climate analysis depends upon the quality of data. The globe covers around 70% by oceans. That means the accuracy of ocean data play vital role at global level studies. In the starting, ships provided ocean temperatures on its route using buckets, later engine room inlet temperature and from 1990 installed buoys. That means prior to 1990 the quality of data was poor that too measurements were taken at very sparse network. IPCC noted that global warming started from 1951. That means between 1951 and 1990, we don't have quality data.

A letter dated 25th January 2016 has been sent by 300 scientists to Lamar Smith, Chairman of the house science committee stating that "We want NOAA to adhere to law of the Data Quality Act". They noted the fact that "As has been acknowledged by numerous scientists, the engine intake data are clearly contaminated by heat conduction from the structure, and as such, never intended for scientific use. Adjusting good data [buoys] upward to match bad data [engine intake] seems questionable.

This is what the NOAA did in erasing the pause observed in the global average temperature anomaly data for the last around 19 years.

This process of adjustment upwards in ocean temperature not only erases the pause but this will modify drastically and thus mislead the natural variability component of global average temperature anomaly and oscillations in different oceans. This will be a dangerous scenario. This way they manipulate to show higher trend. The New Zealand Climate Science Coalition [25$^{th}$ November 2009] presented temperature data for 7 New Zealand stations with longest records [starting around 1850s] before and after adjustments. The increase after adjustments at the 7 stations ranged between +0.40 and +0.89 and -0.15 °C.

On the land surface by around 17$^{th}$ century meteorological stations were started and instrumental measurements were started. In order to understand the cyclic and trend part in global average temperature, we needed quality temperature data as this anomaly is very small compared to the mean temperature. However, the basic problem here is the global average temperature data series were built using poor quality data series from oceans and from land surface. Before 1950 the network was sparse; and even after that the network covers only around 20 to 25% of the globe. For the rest of the Regions the data series were either interpolated and or extrapolated with grid formations.

Ocean temperature data are influenced by several factors associated with general circulation patterns, natural variability factors such as AMO, PDO, IOD, ENSO, Gulf Stream, Humboldt current, etc, etc. They are ocean specific systems. On the land surface temperature is influenced by latitude, longitude, elevation, seasons along with several other localized factors such as micro-climate, heat-island effect in urban areas, cold-island effect in rural areas, advection factor [heat & cold waves, cold & warm breezes], elevation-vegetation, deforestation-reforestation, mining-roads, precipitation [space & time], tides [cyclonic activity & phases of the Moon], filth covering land & oceans after industrialization, etc. All these factors were rarely accounted while adjusting or estimating the data at grid intervals.

The met network is concentrated around urban areas. With the passing of time the changes in land & water use and land & water cover created urban-heat-island and rural-cold-island effects.

Thus, dense met network in urban areas and sparse network in rural areas created biased overestimation of global average temperature anomalies.

In temperature averaging several factors are influencing. Under climate system --natural variability and general circulation patterns over different parts -- is the bottleneck. Because of this every now and then agencies that are controlling global temperature data were manipulated to show what they want or what the half-baked models say. They lowered the past data and moved upwards the current data and thereby raised the mean trend in the anomaly to meet the model predictions [**Figures 1.9 & 2.3**]. Recently to remove hiatus in global temperature data series, the ocean temperature data was manipulated. They proposed three widely differing reasons for such corrections to data series even though they were continuously checked before putting them in to data bank. Even in the case of Satellite data, they initially released the data [this I put in my book of 2008] and later withdrew the same, as this data series trend was far below the ground based data series trend. After manipulating this data they again released. However, even with this, the current satellites and balloons data jointly show [1951 to 2014] around one-third of the global temperature pattern built by them using ground based data [**Figure 1.7**]. The satellites and balloons data took in to account the cold and warm island effects and as well other ecological changes and data gaps. This is balanced global data and yet this has two components: trend and 60-year cycle similar to the global temperature data built based on poor quality ground data. Here the 60-year cycle is natural variability. It only presents the average condition. To fit in to the erroneous ground based data series several qualitative theories were put forward to predict the future rise in the global average temperature anomaly.

### 1.3.3 Sun Related Issues

Solar radiation powers the climate system. There are three basic fundamental ways to change the radiation balance of the Earth, namely change in incoming solar radiation from the Sun, the fraction of solar radiation that is reflected back [albedo] and altering of the long wave radiation from the Earth back towards the space. Climate, in turn, responds directly to such changes, as well as indirectly, through a variety of feedback mechanisms.

Solar radiation is a general term for the electromagnetic radiation emitted by the Sun. The Sun emits radiation [energy] in different wavelengths/bands, starting from X-rays [lower bandwidth side], $\gamma$-rays, ultra-violet, visible, infrared [higher bandwidth side] -- see **Figure 1.4.** The wave length in which maximum energy is emitted depends upon the temperature of the emitting surface. The Sun's surface temperature is around 6000 °C and the Earth's surface temperature is around 10 °C. According to Plank's law the wave length at which the maximum energy is emitted by the Sun is around 0.50µ and the Earth is around 10.0µ. The former is known as short wave radiation and the later is known as long wave radiation. On the top of the atmosphere the energy distribution and day lengths around the globe is presented in **Figures 1.10a & b** – 'a' presents the seasonal and latitudinal daily solar radiation at the top of the atmosphere both in the Northern and Southern Hemispheres; with 'b' presenting day length in a year, that varies between 0 to 24 hours.

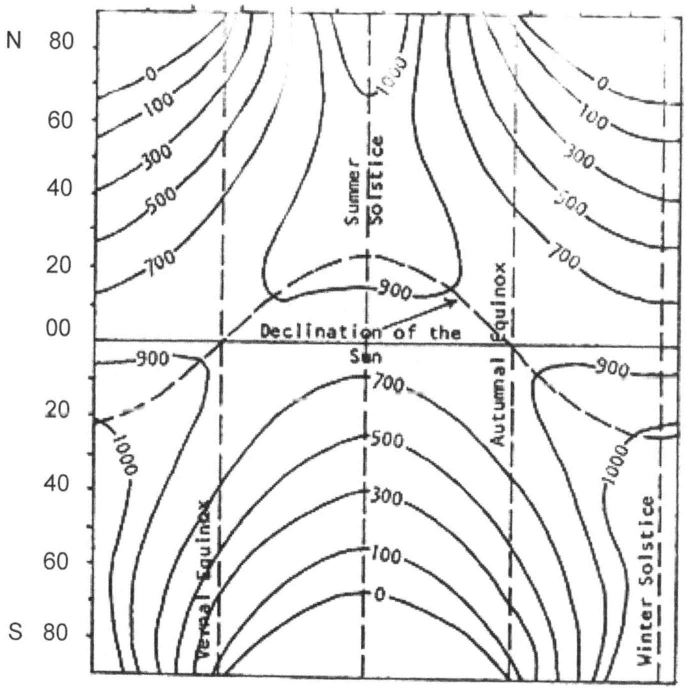


[a]

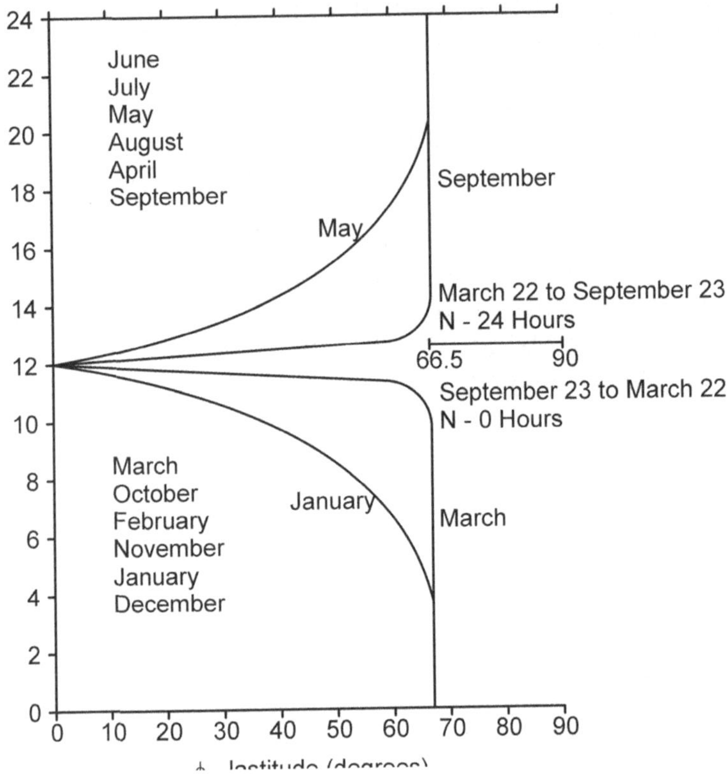


**[b]**

**Figure 1.10** [a] Total daily Solar Radiation at the top of the Atmosphere & [b] Day length

**Note:** Tampering with the surface temperature records is immediately apparent if one plots NH and SH products in, e.g. HADCRUT4:

For almost all of the record, SH and NH move in lockstep. Then, suddenly at the latest turn of the century, NH diverges from SH. SH stays true to the satellite data, independently confirming that the NH in particular has been corrupted to sustain the warming meme.

The Sun's energy while passing through the different layers of the atmosphere under goes changes and after reaching the Earth's surface, it further under goes changes. These can be seen from **Figure 1.11,** a schematic diagram of the Earth's energy balance. About 30% of the sunlight that reaches the top of the atmosphere is reflected back to space. Roughly two-thirds of this reflectivity is due to clouds and small particles in the atmosphere known as 'aerosols'. Light-coloured areas of the Earth's surface – mainly snow, ice and deserts – reflect the remaining one-third of the sunlight. Block bodies reflects less but absorbs maximum and later also emits more while at the same time while bodies reflect more and absorbs less with less emissions.

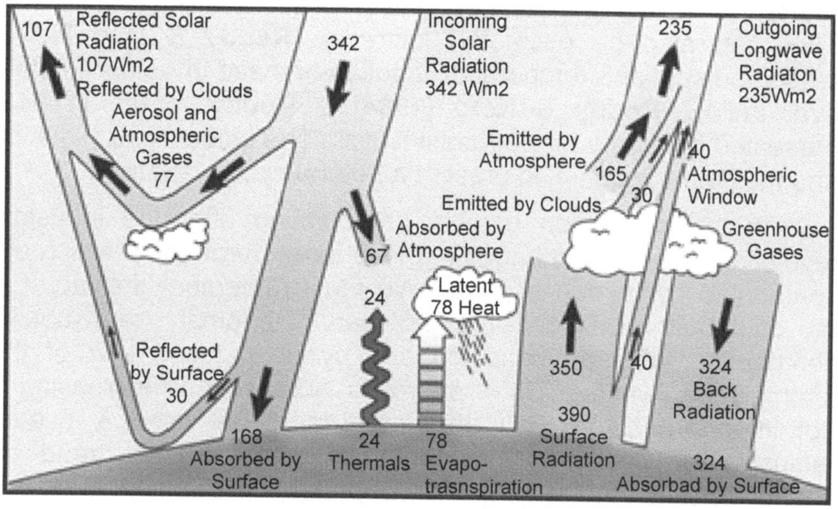


**Figure 1.11** Schematic diagram of Earth's Energy Balance

The energy that is not reflected back to space is absorbed by the Earth's surface and atmosphere. To balance the incoming energy, the Earth itself radiates back some energy to space. The Earth does this by emitting long wave radiation. The reason for the Earth's surface is this warm is the presence of greenhouse gases, which act as a partial blanket for the long wave radiation coming from the surface. This blanketing is known as the natural greenhouse effect. The most important greenhouse gases are water vapour and carbon dioxide. The two most abundant constituents of the atmosphere – nitrogen and oxygen – have no such effect. Clouds, on the other hand, do exert a blanketing effect similar to that of the greenhouse gases; however, this effect is offset by their reflectivity, such that on an average, clouds tend to have a cooling effect on climate (although locally one can feel the warming effect: cloudy nights tend to remain warmer than clear nights because the clouds radiate long wave energy back down to the surface). In this process climate system plays the vital role. Thus, they define the local temperature conditions.

Global [total] solar radiation presents the energy received at the earth's surface after passing through the atmosphere, net radiation presents the balance of incoming and outgoing energy at the earth's surface and evaporation relates to the temperature component from that energy balance under a given climate system. These are measured at few locations around the world. They can also be estimated indirectly. **Reddy & Rao [1976]** presented spatial & temporal distribution of these three parameters over India. **Reddy & Neto [1984]** & **Reddy, et al. [1984a]** presented the same for northeast Brazil. The global solar radiation and the net radiation also present a natural cyclic variation.

With the population growth, urbanization after the industrial revolution, there are changes in these three parameters which are contributed by human actions. We must remember the fact that the contribution of the Sun is not part of anomaly trend but its contribution is mean condition and cyclic variation part of the anomaly. So, whether Sun's energy is increasing or decreasing is not important to evaluate the trend in global [land & ocean] temperature. Cycles of thousands of years present trend on shorter periods but it is insignificant. Under this scenario, there are two options: [1] get better quality data; and [2] start analyzing the data from single station and demarcate homogeneous zones and present the changes in real term.

Due to the rotation of the Earth, the atmospheric circulation patterns tend to be more east-west than north-south. Embedded in the mid-latitude westerly winds are large-scale weather systems that act to transport heat toward the poles. These weather systems are the familiar migrating low- and high-pressure systems and their associated cold and warm fronts. Because of land-ocean temperature contrasts and obstacles such as mountain ranges and ice sheets, the circulation system's planetary-scale atmospheric waves tend to be geographically anchored by continents and mountains although their amplitude can change with time. Because of the wave patterns, a particularly cold winter over North America may be associated with a particularly warm winter elsewhere in the Hemisphere. Changes in various aspects of the climate system, such as the size of ice sheets, the type and distribution of vegetation or topography/topography or the temperature of the atmosphere or ocean will influence the large-scale circulation features of the atmosphere and oceans.

Orbital cycles vary slowly over tens of thousands of years, and at present are in a very slow cooling trend. The variations in orbital cycles may produce a glacial period about 50,000 years from now.

Though the Sun system is the driving force behind the Earth's temperature, they are modified by different forcings created by climate system. This is the natural variations per se. However, as we change the climate system to meet our modern needs, we are creating a change in climate. The sensitivity factors of driving forces are least understood. Though its contribution to temperature regime as such is small, and yet it needs quantification. But this is plagued by several teeth-ache problems.

## 1.3.4 Greenhouse Effect –Global Warming

### 1.3.4.1 Greenhouse Effect

The greenhouse effect was discovered by Joseph Fourier in 1824 and was first investigated quantitatively by Stante Arrhenius in 1896. It is a process by which absorption and emission of infrared radiation by atmospheric gases warm a planet's lower atmosphere and surface. Existence of the greenhouse effect as such is not disputed. Without the Earth's atmosphere, the Earth's average temperature would be well below the freezing temperature of water. Thus, naturally occurring greenhouse gases have a mean warming effect, without which the Earth would be uninhabitable.

On the Earth, the major greenhouse gases are water vapour, which causes about 36 to 70% of the greenhouse effect (not including clouds); carbon dioxide ($CO_2$), which causes 9 to 26%; methane ($CH_4$), which causes 4 to 9%; and ozone, ($O_3$), which causes 3 to 7%. Clouds also affect the radiation balance through cloud forcings similar to greenhouse gases.

### 1.3.4.2 Greenhouse Gases

WMO Fact Sheet No. 4, August 1989 presented the measurements of some of the greenhouse gases in terms of starting year and number of locations these parameters started measuring. Carbon dioxide measurements started in 1959. According to this fact sheet it is clear that very few stations were measuring changing composition of the atmosphere. By that time no data in tropics and only three stations in the Southern Hemisphere. With such a meager data scientists are presenting beautiful smooth change in carbon dioxide with the time. Some of these are discussed by **Reddy [2008].** Carbon Dioxide presents high seasonal variations. Troposphere ozone is created by chemical reactions from automobiles, power plants and other industrial and commercial source emissions in the presence of sunlight. Besides being a greenhouse gas, ground level ozone can also be a harmful air pollutant at ground level, especially for people with respiratory diseases and children and adults who are active outdoors. Stratosphere ozone depleting substances are also greenhouse gases but with the Montreal protocol these substances were replaced with non-ozone depleting substances. However, some of these are greenhouse gases, namely HFCs, PFCs, HCFCs, etc.

Human activities intensify the blanketing effect through the release of greenhouse gases. For instance, the amount of carbon dioxide in the atmosphere has increased by about 35% in the industrial era, and this increase is known to be due to human activities, primarily the combustion of fossil fuels [and removal of forests]. According to work published in 2007, the concentrations of $CO_2$ and methane have increased by 36% and 148% respectively since 1750. However, there is a big question on the quality of this data!!!

Emission scenarios, combined with modeling of the carbon cycle, have been used to produce estimates of how atmospheric concentrations of greenhouse gases might change in the future. Using the six IPCC SRES "marker" scenarios, models suggest that by the year 2100, the atmospheric concentration of $CO_2$ could range between 541 and 970 ppm. This is 90–250% above the concentration in the year 1750. The range shows very poor prediction.

$CO_2$ emissions are continuing to rise due to the burning of fossil fuels and land-use change [a main component of ecological changes]. Emissions can be attributed to different regions. Estimates of changes in future emission levels of greenhouse gases have been projected that depend upon certain economic, sociological, technological, and natural developments. In most scenarios, emissions continue to rise over the century, while in a few, emissions are reduced. Fossil fuel reserves are abundant, and will not limit carbon emissions in the 21st century.

The theory is that these gases trap heat and cause the planet to warm through the process of greenhouse effect. Since 1751 have been released into the atmosphere from the burning of fossil fuels and cement production, increasing atmospheric $CO_2$ from the pre-industrial level of about 280 ppm (parts per million), to a high of 400 ppm in 2013 – we must remember the fact that measurements were very few prior to 1959 and thus they represent heuristic estimates through indirect methods. Methane, which is increasing in the atmosphere due to agriculture and fossil fuel production, traps 84 times as much heat as $CO_2$ for the first 20 years it is in the atmosphere, and is responsible for about one-fifth of global warming since 1750. Nitrous oxide, primarily released through agricultural practices, traps 300 times as much heat as $CO_2$. Over the 20th century, the concentrations of $CO_2$, $CH_4$, and $NO_2$ increased in the atmosphere. But, all these are qualitative estimates.

Here we are forgetting the fact that whether the natural greenhouse gases reached a saturation point to convert the energy in to heat or not. Even the computer models do show the plateau before 2100. That means from the starting point of the plateau extra greenhouse gases over and above the natural greenhouse gases in the atmosphere may not have any influence on the greenhouse effect process. The energy to be trapped and converted is not unlimited. It has a cap defined by the Sun's surface temperature through Plank' Law. Rising levels of atmospheric $CO_2$ do not necessarily cause global warming, which contradicts the core thesis of human-caused climate change, namely global warming.

The Earth's climate record shows that warming has preceded, not followed, a rise in $CO_2$ [**Figure 1.12**]. According to a 2003 study published in *Science*, measurements of ice core samples show that over the last four climactic cycles (past 240,000 years), periods of natural global warming preceded global increases in $CO_2$. In 2010 the *Proceedings of the National Academy of Sciences* published a study of the earth's climate 460-445 million years ago which found that an intense period of glaciations, not warming, occurred when $CO_2$ levels were 5 times higher than they are today. According to ecologist and former Director of Greenpeace International Patrick Moore, Ph.D, "there is some correlation, but little evidence, to support a direct causal relationship between $CO_2$ and global temperature through the millennia."

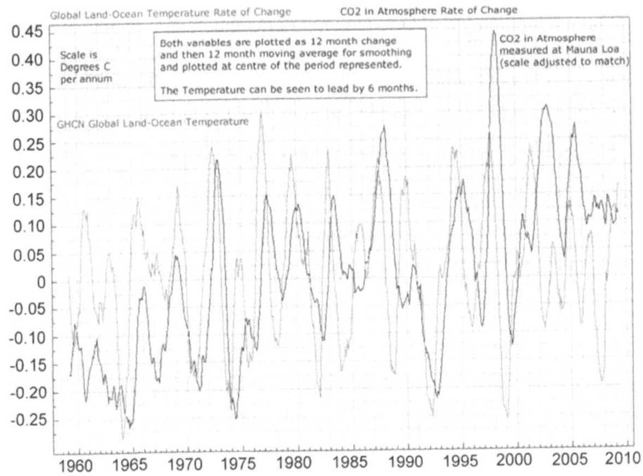


[a]

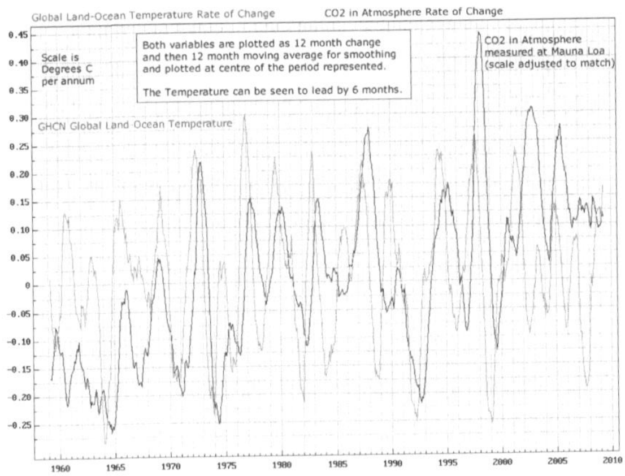


[b]

**Figure 1.12** Historical variation of Temperature versus Carbon Dioxide: [a] Historical pattern and [b] Current pattern

Coming back to the relationship between temperature and $CO_2$ in the modern era we can see that temperature has risen at a steady slow rate in Central England since 1700 while human $CO_2$ emissions were not relevant until 1850 and then began an exponential rise after 1950. This is not indicative of a direct causal relationship between the two. After freezing over regularly during the Little Ice Age the River Thames froze for the last time in 1814, as the Earth moved into what might be called the Modern Warm Period.

There was a 30-year period of warming from 1910-1940, then a cooling from 1940 to 1970, just as $CO_2$ emissions began to rise exponentially, and then a 30-year warming from 1970-2000 that was very similar in duration and temperature rise to the rise from 1910-1940. One may then ask what caused the increase in temperature from 1910-1940 if it was not human emissions? And if it was natural factors how do we know that the same natural factors were not responsible for the rise between 1970 and 2000. You don't need to go back millions of years to find the logical fallacy in the IPCC's certainty that we are the villains in the piece.

That means, though the $CO_2$ was increasing at exponential rate since 1950 [IPCC used this as a starting year of global warming], prior to this and after to this there are plateaus in the global average temperature anomaly patterns. For the last 19 years plateau is the latest.

Patrick Moore noted that $CO_2$ is the currency of life and the most important building block for all life on the Earth. All life is carbon-based, including our own. Surely the carbon cycle and its central role in the creation of life should be taught to our children rather than the demonization of $CO_2$, that carbon is a pollutant that threatens the continuation of life. We know for a fact that $CO_2$ is essential for life and that it must be at a certain level in the atmosphere for the survival of plants, which are the primary food for all the other species alive today.

Human-produced $CO_2$ is re-absorbed by oceans, forests, and other "carbon sinks," negating any climate changes. According to a 2011 study published in the *Asia-Pacific Journal of Atmospheric Science*, many climate models that predict additional global warming to occur from $CO_2$ emissions "exaggerate positive feedbacks and even show positive feedbacks when actual feedbacks are negative." About 50% of the $CO_2$ released by the burning of fossil fuels and other human activities has already been re-absorbed by the earth's carbon sinks. From 2002-2011, 26% of human-caused $CO_2$ emissions were absorbed specifically by the world's oceans. A 2010 study published in the *Proceedings of the*

*National Academy of Sciences* found evidence that forests are increasing their growth rates in response to elevated levels of $CO_2$, which will in turn, lower atmospheric $CO_2$ levels in a negative feedback. According to an August 2012 study in *Nature*, the rate of global carbon uptake by the earth's carbon sinks, such as its forests and oceans, doubled from 1960-2010 and continues to increase.

$CO_2$ is already saturated in earth's atmosphere, and more $CO_2$, manmade or natural, will have little impact on climate. As $CO_2$ levels in the atmosphere rise, the amount of additional warming caused by the increased concentration becomes less and less pronounced. According to Senate testimony by William Happer, PhD, Professor of Physics at Princeton University, "additional increments of $CO_2$ will cause relatively less direct warming because we already have so much $CO_2$ in the atmosphere that it has blocked most of the infrared radiation that it can. The technical jargon for this is that the $CO_2$ absorption band is nearly 'saturated' at current $CO_2$ levels." According to the Heartland Institute's 2013 Nongovernmental International Panel on Climate Change (NIPCC) report, "it is likely rising atmospheric $CO_2$ concentrations will have little impact on future climate."

In nature, there is a limit for everything. For example, seed has a limit for production under certain inputs conditions. By adding more and more inputs there will be no increase in productivity and thus present a plateau with any further increasing levels of inputs. $CO_2$ in nature is variable with the time and even under this variability it optimizes the use of energy in its wavelength bands. By increasing anthropogenic greenhouse gases has little impact as the energy emitted by the Sun as it has a cap.

Though water vapour component of greenhouse gases might have been come down with deforestation during the hunter to domestication process, this was more than enough was compensated with development of water resources and irrigated agriculture. However, seasonal and annual variations along with the natural variability the greenhouse effect changes and so also temperature and other meteorological parameters.

Under these scenarios, UN must look at reducing pollution through proper technology changes rather than on harping on anthropogenic greenhouse gases emission reduction.

### 1.3.4.3 Aerosols

The most dramatic change in aerosol-produced reflectivity comes when major volcanic eruptions eject material very high into the atmosphere. Rain typically clears aerosols out of the atmosphere in a week or two [only if there is a rain, otherwise it will stay in the atmosphere for more time], but when material from a violent volcanic eruption is projected far above the highest cloud, these aerosols typically influence the climate for about a year or two before falling into the troposphere and being carried to the surface by precipitation. Major volcanic eruptions can thus cause a drop in mean global surface temperature of about half a degree Celsius that can last for months or even years. Some man-made aerosols also significantly reflect sunlight. All these, however, are too generalized inferences as there are several complications are involved at a given place and in a given time of the season.

A gradual reduction in the amount of global direct irradiance at the Earth's surface was observed from 1961 until at least 1990. This is known as dimming effect. Solid and liquid particles known as *aerosols*, produced by volcanoes and human-made pollutants, are thought to be the main cause of this dimming. They exert a cooling effect by increasing the reflection of incoming sunlight. Radiative forcing due to aerosols is temporally limited due to the processes that remove aerosols from the atmosphere. Removal of aerosols by clouds and precipitation gives the Troposphere aerosols an atmospheric life time of only about a week, while the Stratosphere aerosols can remain for a few years.

In addition to their direct effect by scattering and absorbing solar radiation, aerosols have indirect effects on the Earth's radiation budget. Sulfate aerosols act as cloud condensation nuclei and thus lead to clouds that have more and smaller cloud droplets. These clouds reflect solar radiation more efficiently than clouds with fewer and larger droplets, a phenomenon known as the Tworney effect. This effect also causes droplets to be of more uniform size, which reduces growth of raindrops and makes the cloud more reflective to incoming sunlight, known as the Albrecht effect. Indirect effects are most noticeable in marine Stratiform Clouds, and have very little radiative effect on convective clouds. Indirect effects of aerosols represent the largest uncertainty in radiative forcing.

Soot may either cool or warm the Earth's climate system, depending on whether it is airborne or deposited. Atmospheric soot directly absorbs solar radiation, which heats the atmosphere and cools the surface. When deposited, especially on glaciers or on ice in arctic regions, the lower surface albedo can also directly heat the surface. The influences of atmospheric particles, including black carbon, are most pronounced in the tropics and sub-tropics, particularly in Asia.

Variations of Volcanic aerosols and possibly solar luminosity appear to be primarily causes of observed fluctuations about the mean trend of increasing temperature. Volcanoes aerosol impact is not new. This not only creating cooling effect but it reduces the energy reaching the Earth and thus energy available for greenhouse effect for global warming. To create greenhouse effect, we need energy also, and if this energy is reduced drastically then the global warming is automatically comes down.

Reports suggest that the violent volcanic eruptions would cause some warming for the first few months, and then, when the larger particles have settled, larger global cooling by 1 – 2 $^{o}$K. The degree of this change depends upon the season in which this has occurred. For example: Eruption of Mt Pinatubo in the Philippines in June 1991, there was a slight global warming for the first 3 months after the eruption, possible due to the coincident appearance of an of El Niño event. After the removal the effect of El Niño Southern Oscillation, a fall of global surface temperatures by about 1.0 $^{o}$K in the first year was observed. The anomaly fell to about 0.5 $^{o}$K after 16 months, and erratically returned to normal over the next three years. However, the lower stratospheric temperature in the Tropics rose by over 1.5 $^{o}$K within 3 months after the eruption, as a result of the dust's absorption of solar radiation. Then the temperature there fell back to normal over the following 18 months, as the dust settled out. So, they are also season and region specific depending up on where they occur. Similarly earthquakes, wars, nuclear tests/nuclear dust, forest fires, etc also contribute to changes in weather, a short term change. However, ecological changes component will not be affected significantly by these processes.

So this is short lived change. This comes under irregular variation component of systematic variation. They form part of intra-seasonal and intra-annual variability of weather. **Reddy,**

**et al. [1976]** studied the Power spectral analysis of Lower Stratospheric Meteorological data of H, T, u & v" for 11 stations between 1 to 13 degrees latitude belt and 2 to 103 degrees longitude belt for 100, 50, 30 mb levels in India. The inferences concurs those findings wherein the cycles varied between 6 months to 36 months.

Here the most important point we are missing is the reduction in relative energy available to greenhouse effect and thus lowering the global warming component automatically without affecting the ecological changes component.

### 1.3.4.4 Ozone Versus Carbon Dioxide

**Figures 1.12a & b** presents the relationship between carbon dioxide with temperature: [a] presents the Historical pattern & [b] presents the Current pattern. This is an anti-thesis of global warming phenomenon. Here the carbon dioxide is the after effect with the temperature. *In a book:* "What Really Causes Global Warming?" and "Greenhouse Gases or Ozone Depletion?" Peter Langdon Ward observed that "t*wo different styles of volcanic eruption appear to have been the principal determinants of climate change throughout geologic time.* In brief, we find that major temperature changes throughout Phanerozoic time can be fully explained with two different styles of volcanic eruptions: explosive volcanism causing global cooling and effusive volcanism causing global warming. It is well-known that aerosols from explosive volcanoes, such as the 1991 eruption of Pinatubo, reflect and scatter sunlight, causing global cooling. What we found is that all volcanoes emit chlorine and bromine, which are observed to deplete the ozone layer, allowing increased irradiance of Earth by solar UV-B radiation, causing global warming. UV-B is 48 times more energy-rich than Earth's IR radiation absorbed by carbon dioxide. Under conditions normal before 1965, ultraviolet-C (UV-C) warmed the upper atmosphere, UV-B primarily warmed the ozone layer, UV-A and visible light warmed Earth. CFCs, when they rise to the level of very cold polar stratospheric clouds (PSCs), release chlorine that depletes ozone, causing more UV-B than usual to reach Earth's surface, thus cooling the ozone layer and warming Earth. Effusive volcanoes emit chlorine and bromine, which deplete ozone, leading to global warming. Explosive volcanoes similarly deplete ozone, but also eject megatons of water and sulfur dioxide into the lower stratosphere, forming globe-encircling aerosols whose molecules soon grow large enough to reflect and scatter sunlight, causing net global cooling. These are explained in **Figures 1.13 & 1.14**.

**Figure 1.13** presents a seasonal march of temperature with carbon dioxide and ozone. The figure presents the mean monthly values of northern hemisphere temperature anomalies and ozone depletion anomalies for the period 1975 to 1998 and of atmospheric carbon dioxide concentrations at Mauna Loa, Hawaii, since 1961, normalized as percentages. Carbon dioxide values, peaking in May, show only a minor effect on temperature anomalies, but coincidence of the peaks in ozone depletion and temperature in March suggest a possible causal relationship.

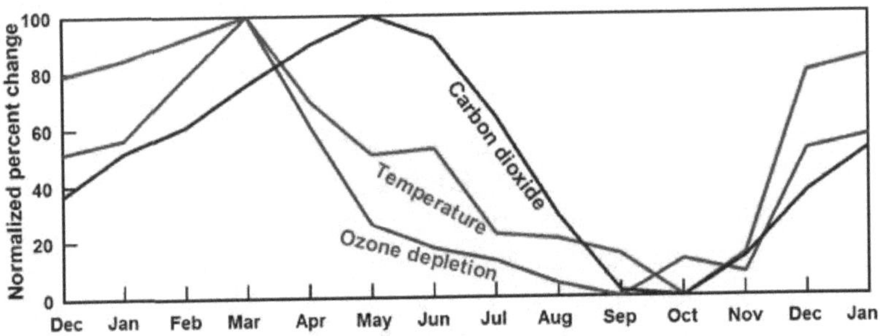


The relationship of Ozone Depletion to Temperature

**Figure 1.13** The relationship of Ozone Depletion & Carbon Dioxide to Temperature

**Figure I.14** presents the trends in Temperature, Troposphere chlorine and ozone depletion over the past 70 years are distinctly different from trends in concentrations of greenhouse gases such as carbon dioxide. Ocean heat content increased with increasing ozone depletion and continues to increase while ozone depletion remains greater than levels prior to 1970. Carbon dioxide levels appear related to ocean heat content through the solubility of $CO_2$ as a function of water temperature.

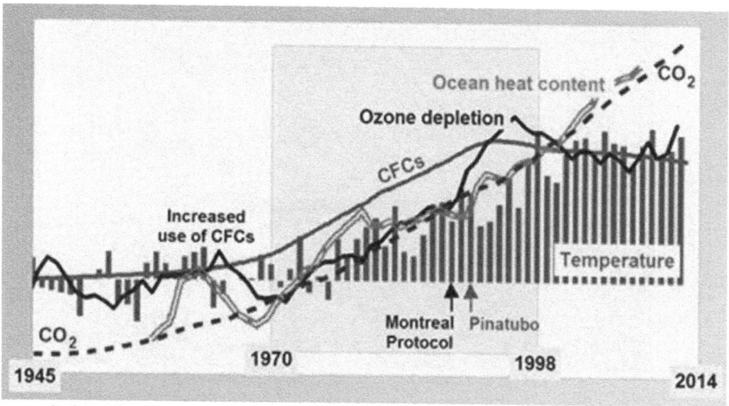


**Figure 1.14** Trends in temperature versus Ozone and Carbon Dioxide

### 1.3.4.5 Oceans Covered by Filth

Earth's surface is being noticeably altered by the production of long-lasting man-made materials, resulting in us entering an 'age of plastics'. Over planet's oceans and lands will be buried by increasing layers of plastic waste. They are inert and hard to degrade.

Though there is no scientifically sound estimates exists for the size or mass of the filth entering the Oceans & water bodies, atmosphere & land areas but these significantly influence the incoming and outgoing radiation and thus temperature regime. This will go on increasing with the growth of population. **Figures 1.15a & b** present Great Pacific garbage patches of plastics.

[a]

[b]

**Figure 1.15a & b** Garbage patches in Great Pacific

On Oceans, cruise ships generate an astonishing amount of pollution: up to around 25,000 gallons of sewage from toilets and 143,000 gallons of sewage from sinks, galleys and showers each day. Coastal environment and marine life are at risk from the threats of bacteria, pathogens and heavy metals generated in these waste streams. Ships generate 15 to 30% of the World's smog-forming emissions. Bunker fuel burned by ships is 1,000 times dirtier than highly diesel used by trucks and buses. The shipping industry burns 300 million tons of bunker fuel per year. A single ship coming in to harbor produces the smog-forming emissions of 350,000 new cars. Ships also generate huge amounts of solid waste. Ocean waters are also affected by oil spills. Polluted waters are also entering oceans/rivers from agriculture rain water runoffs, domestic sewages, etc.

Also with the plastic became part of the life, are covering every bit of open land in and around urban areas and now it is also spread to rural villages. Air traffic emissions also covered upper atmosphere that significantly influencing the radiation balance and thus changes in temperature and other met parameters.

We are not accounting all these factors that significantly influencing the radiation balance and thus influencing the temperature and other met parameters. With the spread of filth over sea/ocean waters, the Sun's energy absorbed and reflected by water and in-turn energy emitted by water will be changing with the increasing levels of filth over the water. It may impact the general circulation in sea/ocean waters. Commercial jet aircrafts also fill the atmosphere with certain other types of filth.

**Urban Filth:** At present around 30% of the population are living in urban areas under unhygienic conditions in several parts of the globe. The population may reach 50 to 60% of the total by 2050. They generate huge quality of filth as the sustainable management of solid water management in developing countries is a myth. There are several types of wastes, namely industrial, biomedical, E-waste, plastic waste, batteries waste, mining/construction waste and municipal solid waste, etc. Solid waste management is given a very low priority in developing countries. Dumping them indiscriminately and burning them are the primary factors that affect local weather conditions in terms of temperature. We rarely take in to account these factors which in fact will increase with the time covering wide areas on the ground like urban concrete jungle.

### 1.3.4.6 Global Warming

From the definition of IPCC that the increased global average temperature anomaly since 1951 has two parts, namely (1) one caused by the greenhouse effect and (2) the other caused by non-greenhouse effect. The former's contribution to the global average temperature since 1951 is extremely likely that more than half; and thus the latter's contribution is less than half. Also, at the same time the anthropogenic greenhouse effect component has two components, namely (1a) the anthropogenic increase in greenhouse gas concentrations and (1b) the other anthropogenic forcings. However, IPCC is not sure of the quantitative contribution of these three groups on the global temperature.

Sophie Lewis, from the Australian Research Council's Centre for Excellence in climate science, dissecting statements about climate change records made by the former Prime Minister of Australia, Tony Abbott in 2013, observed [22$^{nd}$ January 2016] that "To test if that might be the case, Lewis ran a series of climate models in which the greenhouse effect was removed – so all that was left was natural variability. Unsurprisingly, in those models, high temperature records were less common than they are in reality. In other words, the record-breaking that we have seen cannot be explained by natural variation." -- [1] IPCC noted more than the half of global average temperature anomaly is associated with the greenhouse effect component [this includes human component namely global warming and non-human component related to volcanoes and other aerosols component] and the rest to non-greenhouse component [mainly human induced component]. That means even after removing the global warming component, still the other two components along with the systematic variations components [fluctuations & irregular variations]. The so called models in fact haven't taken these in to account. The difference between 2014 and 2015 is 0.18 °C [but they are different under different data sets and under satellite data sets]. The predicted rise associated El Nino is about 3 °C. With every peak of El Nino, on either side two depressions will occur and with this 2016/17 will automatically follow the hiatus.

Unless these three parts are quantified, we cannot say it is settled science as in the three parts, (2) and 1(b) are not global in nature or character; but they are only either local and or regional in nature or character. Thus, this has no role at global scale except going in to the global averaging of temperature. With such qualitative groupings how can, IPCC argues it is a settled science, a big question?

IPCC is arguing on 2 °C by the end of the century. The question is, is this relates to all the three parts or only one part that relates to the anthropogenic greenhouse gases concentration [1(a)]? IPCC needs to clarify this before asking individual governments to bring down anthropogenic greenhouse gases or before some governments imposing taxes on emissions. Also, unless this is clarified by IPCC, how can we quantify its impacts on nature? First let us get clear idea on the subject. IPCC was cautious not to refer '1a' as global warming and instead uses climate change!!!

There are few other issues that need to be considered like filth covering the oceans, filling the atmosphere with filth generated by commercial jet aircrafts other than volcanic aerosols, etc.

Whether we get an answer to such issues or not, India must look into the components of climate change that are local or regional in nature that directly influence the agriculture and water resources, instead of running after emissions and $100 billion dollars.

**Sensitivity factor:** Volumes and volumes were written by thousands of people on Radiative Forcings [RF]; but nobody gave quantitative figures for anthropogenic greenhouse effect part, which we term it as global warming. Even IPCC made qualitative statements like tossing a coin: you may get head or tail only. Now, let us come to IPCC – they monotonically reduced the sensitivity factor in carbon dioxide forcing from SER to AR5 [in AR4 it was 1.95 and in AR5 it was 1.55]. Also, in AR5 they qualitatively divided global temperature in to two parts, namely *more than half of the observed increase in global average* surface temperature from 1951 to 2010 was caused by the *anthropogenic increase in greenhouse gas concentrations* and other *anthropogenic forcings* together. This relates to the trend component only. Cyclic variation component is form part of systematic variation [60-year cycle – presented in my 2008 book from my earlier analysis] superposed on the trend. Anthropogenic component has two parts, namely one is associated with greenhouse effect and the other is non-greenhouse effect component. If we divide the 50% [though IPCC said more than 50%, even 50.1% is more than 50% only] equally in to these two components, the greenhouse effect component [global warming component] is around 25%. This is the component associated with anthropogenic greenhouse gases [carbon dioxide] increase in the atmosphere. The second 25% component is not associated with carbon dioxide increase in the atmosphere but relates to non-human factors. The remaining around 50% of the global temperature raise is due to non-anthropogenic component. The net, non-anthropogenic greenhouse gases component is around 75%. This part not only contributes to the changes in temperature but also to the changes in other meteorological parameters directly and indirectly. This is a local and regional phenomenon will not contribute to global phenomena. According to this, if the global temperature raise for example is 1.0 °C, the global warming component is 0.25 °C. If we take the satellite and balloon data, 1.0 °C is replaced by 0.35 °C only. Here, the major advection factor was not taken in to account at local/regional level. IPCC must come up with the quantitative percent – 25% or less or more then only there will be sense in it.

According to NASA/Goddard Space Flight Centre new study published December 14, 2015 in the Journal "Nature Climate Change", to understand climate change, researchers need to know the Transient Climate Response (TCR) and Equilibrium Climate Sensitivity (ECS) of Earth. Both values are projected global mean surface temperature changes in response to doubled atmospheric carbon dioxide concentrations but on different time scales. TCR is characteristic of short-term predictions, up to a century out, while ECS looks centuries further in to future, when the entire climate system has reached equilibrium and temperature have stabilized.

There have been many attempts to determine TCR and ECS values as the history of temperature changes over the last 150 years and the measurements of important climate drivers such as carbon dioxide. As part of that calculation, researchers have relied on simplifying assumptions when accounting for temperature impacts of climate drivers other than carbon dioxide, such as tiny particles in the atmosphere known as aerosols such as those emitted in volcanic eruptions act to cool Earth, at least temporarily, by reflecting solar radiation away from the planet. In a similar fashion, land use changes such as deforestation in Northern Latitudes result in bare land that increases reflected sunlight. But the assumptions made to account for these drivers are too simplistic and result in inaccurate estimates of TCR and ECS, said Gavin Schmidt, co-author of the study. He further said that "The problem with that approach is that it falls way short of capturing individual regional impacts of each of those variables".

This means that Earth's climate sensitivity to carbon dioxide or atmospheric carbon dioxide's capacity to affect temperature change – has been underestimated. The result dovetails with a GISS study published last year that puts the TCR value at 1.7 °C; the IPCC, which draws its TCR estimate from earlier research, places the estimate at 1.0 °C. Schmidt said that "If you have got a systematic understanding of what the greenhouse gases driven change would be, then you systematically understand what's going to happen in the future when greenhouse gases are by far the dominant climate driver". However, this news release and the abstract do not say what their new climate sensitivity estimates are. The supplemented [http://www.nature.com/nclimate/journal /vaop/ncurrent/nclimate2888-s1.pdf] says TCR is 1.4 °C and ECS is 2.3 °C.

The authors talked of aerosols of volcanic activity but there are several types of aerosols for the last centuries related to wars, dust storms, mining activities, building activities, etc. Where carbon dioxide is not uniformly spread throughout the globe, how you can expect it produce uniform response? But some argue that even with 1000 ppmv for the first 1000 m the net effect would be no more than 0.1 °C extra warming. How, they arrived at this value is surprising. If they are claiming 1.7 °C now, how in they have been trying to hang on to 3-4 °C is another big question – though some argued that this is ECS. Also, the natural climate response to a change in any one of the components of the water cycle is not a part of the estimate and yet Trenberth insists the missing heat is hiding in the ocean and Karl insists the pause is due to underestimating of SST. They argued greenhouse gases by far the dominant climate driver, how can it be possible be known now that this will be the case at some point in future? Or what evidence? In fact several studies put the carbon dioxide contribution only 12-14% of the temperature effect then how it becomes dominant factor, is a big question.

In order to test their theories, we have to have accurate measurements of average global temperature anomalies???

### 1.3.5 Non-Greenhouse Effect – Ecological Changes

The main component of human activity is ecological changes, not associated with greenhouse effect [more details can be seen from **Reddy, 2008**]. Ecological changes are associated with changes in land use & land cover and water use & water cover. They include urban-heat-island effects and rural-cold-island effects. Urbanization, deforestation, mining, etc cause the heat island effect. This component is a part of global temperature raise. The quantum relates to the location of met stations. We must remember one thing here, majority of met stations were/are located in urban areas by revenue department for convenience. Because of this urban-heat-island effect dominates the average temperature. Changes in agriculture [dry-land to wetland], water resources development, reforestation, etc will bring down the global temperature if it is added; but in reality the met network was and is sparse in such areas and thus the effect is under reported in averaging of temperature. This is clearly evident from the differences in Northern Hemisphere and Southern Hemisphere temperature patterns [**Figures 2.1 & 2.2**]. However, satellites and balloons data takes these in to account.

## 1.4 Concluding Remarks

Climate is what you expect; weather is what you get. Weather refers to short-term events in the atmosphere, while climate relates to longer time period events in the atmosphere. Every National Meteorological Services [NMS] have published Climate normal books that contain averages and extremes for individual meteorological parameters and meteorological stations. The period used was for 30 years. The current extremes in weather haven't crossed the limits in the normal books world over.

The atmosphere is not an isolated system. It interacts with other components of the Earth system – the Oceans, for example. But it is also in contact with the Cryosphere (ice and snow), the Biosphere (animals and plants), the Pedosphere (soil) and the Lithosphere (rocks). All of these elements together compose the climate system, whose individual components and processes are connected and influence each other in diverse ways. These are further modified by the general circulation systems that include wind & pressure that vary with region to region. It also changes with the climate system.

Climate refers to all meteorological parameters. The met parameters interacts each other in different ways over different parts of the globe within climate system and general circulation patterns. Climate change is, thus, a local-regional specific change and not a change at global scale. Global warming associated with only temperature. There is no warming at global scale but it is a sum and total of positive and negative variations over different locations-regions of the globe.

Climate change consists of systematic variations in which Irregular variations consisting of intra-seasonal, intra-annual variations and fluctuations, known as rhythmic or cyclic variations are part and are also location-region specific; and man-induced variations or associated with human activity in which ecological changes associated with the changes in land & water use and land & water cover changes which is a location-region specific; global warming associated with the anthropogenic greenhouse gases; though greenhouse gases composition is location-region specific but considered it as global factor; and other causes associated with natural and human actions but they are intermittent in nature. Climate change is broader subject relating to met parameters within which global warming is only one component related to temperature only.

The basic problem is people shy of using the word global warming as they are not sure on its magnitude of contribution to global average temperature anomaly and thus instead they use the words such as climate change. While changing the definition of climate change IPCC and UN agencies played game with ecological changes component of human induced changes, which is the major component that is causing the trend in global average temperature. To avoid any ambiguity, it is better to use global warming to refer to change in temperature associated with the increased levels of anthropogenic or human-induced greenhouse gases through greenhouse effect. However, the basic question is, whether this additional greenhouse gases are really converting in to raise the temperature as the amount of greenhouse gases in the nature is more than sufficient to convert energy in to temperature. In nature zero to infinity is not a reality but they are bounded with plateau on either side.

All the light emitted by the surface in the strongest $CO_2$ bands was completely absorbed at pre-industrial levels. That is below 280 ppm. Therefore, the additional $CO_2$ added to the atmosphere through human actions or through any other process may have little contribution. Because of this, to show there is an accelerated increase in global average temperature anomaly, data managers started adjusting the basic temperature data by lowering the past data series and raising the current data series, though contrary is evident in satellite & balloon data series. While adjusting, they forgot the fact that global average temperature anomaly is not contributed by the global warming component alone but it is contributed by several factors, such as human and non-human factors. This process of adjustments upward in ocean temperatures, not only it erased the pause for the last 19 years but also will modify drastically that mislead the natural variability component of global average temperature anomaly and oscillations in different oceans. This will be a dangerous scenario.

Global average temperature is a function of Sun's energy and climate system at any given place. This has intra-annual and intra-seasonal variations. Also they contain both long term and short term decadal oscillations. Over the mean, the anomaly that varies with the time consists of cyclic variation and trend. That is, the global average temperature anomaly consists of natural fluctuation component and natural & human induced trend. Sunspots and global solar radiation and net radiation/radiation balance at the Earth's surface present 11-year cycle and its multiples.

Volcanic activities contribute to natural trend part. The human induced trend part consisting of four components; and is associated with aerosols other than those from volcanic activity, filth over sea/ocean water, greenhouse effect and non-greenhouse effect. There are no quantitative estimates for the four types of components as the world changing with the passing of time; and thus % contributions by these four differ with the time. IPCC has only qualitative estimates for some of these components.

Globally, the temperatures are showing high variations in trend and in average anomalies with country to country, region to region within a country, land to ocean, SH to NH, etc. Also, measuring equipment & met stations location are themselves creating large variations. There is no exception to even the satellites data as manipulation of data becomes a common factor to justify global warming and its impacts. So, we are trying to build castles, known as global warming, based on false foundations. However climate change is a reality.

It is quite surprising to note that the Papel 'Laudato Si' encyclical on global warming has become a source of division and prompted heated exchanges between senior Church figures at a top level meeting in Rome in December 2015. Even Pope Francis in his address on Christmas-eve on 25$^{th}$ December, did not mention a word on the climate change.

Chapter

# 2

# Impacts of Global Warming: Ground Realities

## 2.1 Introduction

Wikipedia talks "The internally driven changes in the oceans such as changes in the Gulf Stream also occur within a time frame of decades or a few centuries. These have a decisive influence on climate and on the concentration of greenhouse gases in the atmosphere because they are strongly involved in global mass cycles such as the carbon cycle. For example, $CO_2$ dissolves easily in water. However, the oceans have taken up about half of all the carbon dioxide produced by the burning of fossil fuels since the beginning of the industrial revolution, which has clearly dominated the natural variations. Whether the climate will change in the future, and by how much, can therefore be also deduced from the oceans.

Climate will change very slowly in the future because the oceans with their immense volumes of water react very gradually to change. Therefore, many but not all of the consequences of climate change triggered by human activity will only gradually become noticeable. Some of these consequences could actually be irreversible when certain thresholds are crossed. At some point it will no longer be possible, for instance, to stop the complete melting of the Greenland ice sheet and the resulting seven-meter rise of sea level. The position of the threshold, however, is not precisely known. But one thing is certain: Even if the emission of carbon dioxide were stabilized at today's levels it would not lead to a stabilization of the carbon dioxide concentration in the atmosphere, because carbon dioxide is extremely long-lived and the carbon dioxide sinks, mainly the oceans, do not absorb it as quickly as we produce it.

The situation is different for short-lived trace gases like methane ($CH_4$). If methane emissions were stabilized at the present level, the methane concentration in the atmosphere would also stabilize, because methane diminishes in the atmosphere at about the same rate as it is emitted. In order to maintain the carbon dioxide concentration at a given level, the emissions have to be reduced to a fraction of the present amounts. Long after the stabilization of carbon dioxide levels, the climate will still further continue to change because of its inertia.

Climate models indicate that the near-surface air temperature will rise for at least a hundred years. Sea level will continue to rise for several centuries because seawater expands slowly as a result of the gradual warming of the deep sea, and because the continental ice sheets in the Arctic and Antarctic will probably react very slowly to the warming of the atmosphere, and the glaciers will continue to melt for many millennia. It will therefore be a long time before sea level achieves a new equilibrium.

But scientists also believe it is possible that, if the warming is strong, the Greenland ice sheet could completely melt within this millennium and disappear into the ocean. The ice sheet could actually break apart and giant pieces fall into the sea. The enormous amounts of fresh water could cause a critical change in ocean circulation, for example, in the Gulf Stream. In an extreme scenario, sea level could rise by more than a meter per century, regionally by even more. The inertia of the climate system and the danger that the trend is irreversible should be sufficient reasons for forward-looking action. One should always keep in mind that the impacts of climate change that are measurable today do not yet reflect the total extent of climate change already caused by humans in the past. Humankind will only begin to feel them sharply in a few decades but has to take action right away."

All these are hypothetical theories but to what extent they are close to reality needs an answer. This is possible only when global warming component is quantified. A report [28$^{th}$ January 2016] state that "There would have to be unprecedented heat in the atmosphere/oceans to produce unprecedented weather, and the Earth's atmosphere is not experiencing unprecedented heat right now, despite what the advocates of the global warming/climate change theory say. When it gets hotter than 1934, was, let me know. There were some pretty severe weather outbreaks worldwide during the very hot decade of the 1930's: lots of droughts; lots of massive heat waves; lots of hurricanes and tornados; and lots of melting glaciers. Kind of like the scenarios the advocates claim we are facing today, in our current environment. But we are not there yet, by any credible measurement. We are not hot enough to duplicate the 1930's weather. How do I know? Well, have you seen "lots of droughts; lots of massive heat waves; lots of hurricanes and tornados; and lots of melting glaciers" worldwide since the beginning of the 21st Century? And humans didn't cause the extreme overheating in the 1930's. It was all natural. People, who assume we are experiencing unprecedented heat in the atmosphere today, are assuming too much. Most of them are assuming too much because they believe the climate science charlatans at NASA, NOAA and the IPCC, who have fooled them into believing a lie by showing them phony, doctored surface temperature charts."

## 2.2 Variations in Temperature

### 2.2.1 Temperature Data

Meteorological stations were established to measure weather data all over the world and these stations have been increasing with the time; in some cases station locations have been changed or they are in bad shape/condition or in a dilapidated condition; or instruments changed; unit of measurements changed – °F to °C and inches to millimeters. Some followed standard Stevenson Screens and others used modern gadgets. Some have data gaps due to disasters or wars or epidemics, etc. Though oceans occupy around 70% of the global area, the meteorological measurements are sparse and that too are highly unevenly distributed in terms of space and time.

Thus, more than 75 to 80% of the global areas have no met stations. As the met stations were maintained by revenue departments, they are mainly located in urban areas for ease of controlling them. Met stations are sparse in rural areas. Since 1979 [though started in 1972] satellite observations are available. Also, Radiosonde Balloon observations are available at few locations around the globe.

In temperature the averaging introduces an error while rounding of the value in the second place of decimal value to the first place of decimal, as the units of measurements are reported to the first place of decimal only. For example: 30.45 °F or °C is taken as 30.5 °F or °C and 30.55 °F or °C is taken as 30.5 °F or °C only. Based on the odd or even number in the first place of decimal, the adjustments are carried out accordingly. The difference between 30.45 and 30.55 is 0.10 and °F is less than that of °C. The rounding impact on °F is different from °C.

John Greenberg [13-02-2015] presented an article "Fox News host: Climate scientists 'fabricated' temperature data" in "PunditFact". The report says: Fox News host Dana Perino joined several of her colleagues this week in casting doubt on the data scientists use to track temperature changes over time [09-02-2015]. Co-host Kimberly Guilfoyle interjected that it was "fraud science" and Perino said, "Yes, I agree." The U.S. Senate may have voted 98-1 that "climate change is real and not a hoax," but the accusation that government scientists have cooked the books and invented a warming trend is as robust as ever. A couple of days before she spoke, the British paper *The Telegraph* carried an opinion piece entitled, "The fiddling with temperature data is the biggest science scandal ever". The article drew on the work of a well-known climate change skeptic, Paul Homewood, who on Feb. 4, 2015, blogged about alterations in the temperature records in Paraguay and some arctic locations. The *Telegraph* article concluded that these changes were part of the "greatest and most costly scare the world has known."

**Figure 2.1** presents [a] Annual mean Temperature departures from 1961-90 normal and linear trends for different regions in Canada, 1948-2009, [b] The quality of temperature sitting versus temperature trend in 9 US climate zones, & [c] 2015 Global Surface Temperature Departures from Normal.

Anthony Watts and his group studied US historical met network and temperature trends. They concluded that NOAA's US climate network shows a lower 30-year temperature trend when high quality temperature stations unperturbed by urbanization are considered. They grouped several types of met stations totaling of 1218 of NOAA into three groups, namely 410 stations as that do not require adjustments [class 1 & 2 -- compliant]; 808 stations as that require adjustments [class 3, 4 & 5 – non-compliant]; and combined 1218 stations [classes 1 to 5] official NOAA adjusted. **Figure 2.1b** presents the trend in these three groups for 1979 to 2008 [30 years] in 9 climate zones. In the first group the trend varied between 0.040 and 0.292; in the second group they are 0.219 to 0.442; and in the third they are 0.224 to 0.409. They concluded that "NOAA's U.S. Climate Network shows a lower 30-year temperature trend when high quality temperature stations unperturbed by urbanization are considered". These stations don't require adjustments to the data.

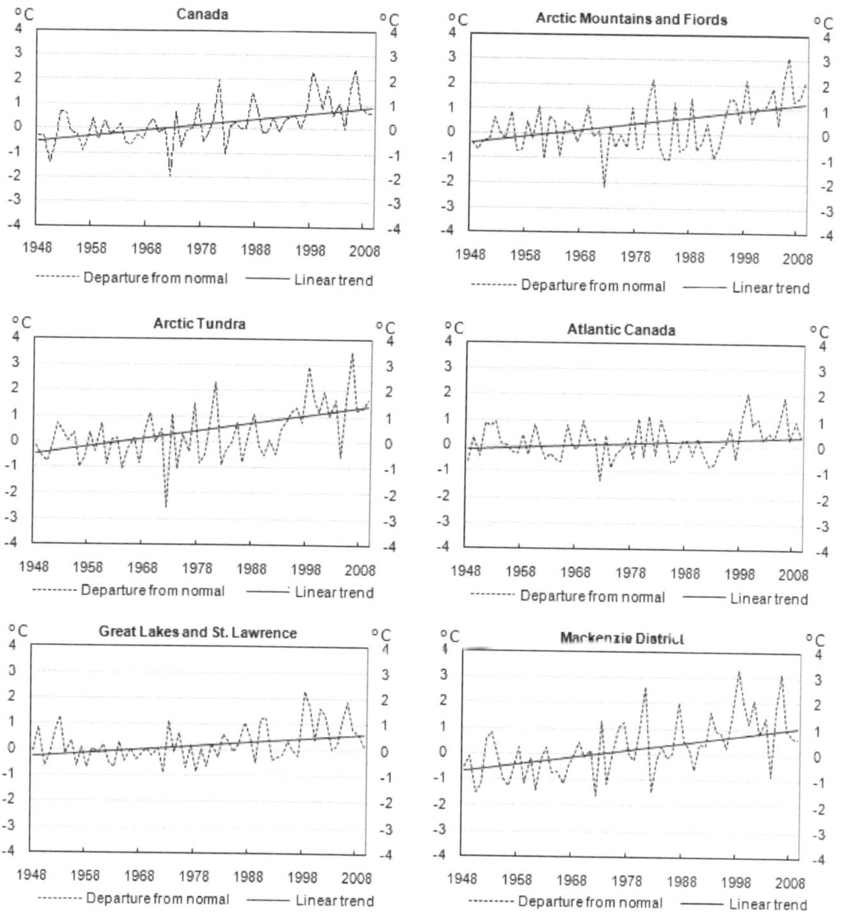

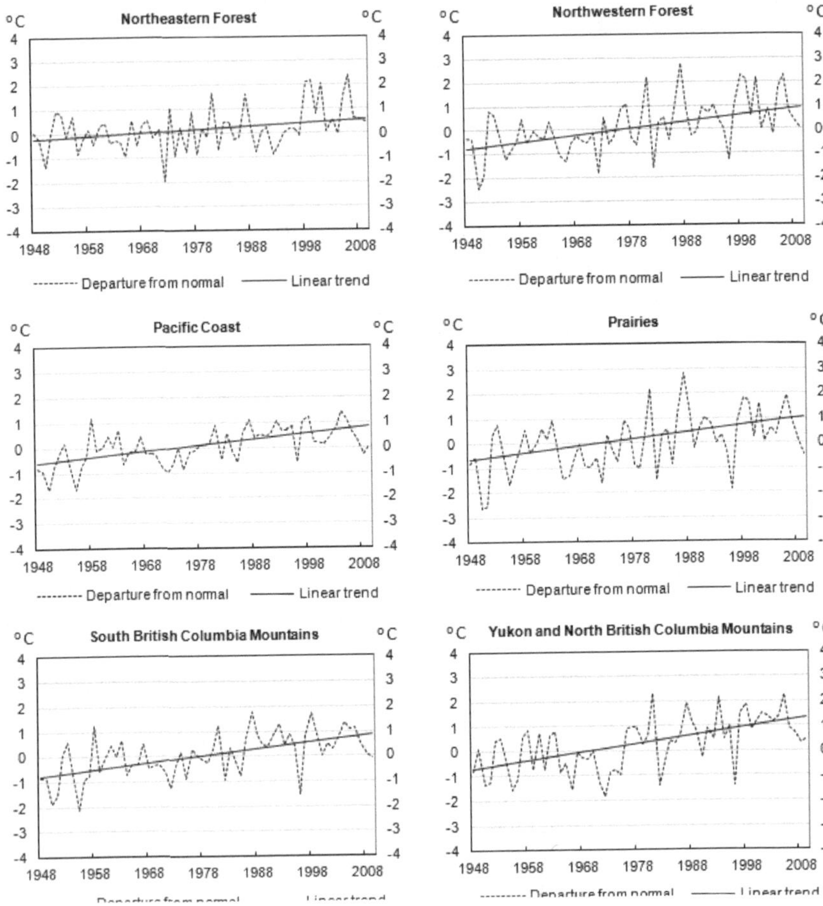
°C
Northeastern Forest
°C
4
3
2
1
0
-1
-2
-3
-4
1948 1958 1968 1978 1988 1998 2008
-------- Departure from normal
——— Linear trend
Northwestern Forest
-------- Departure from normal
——— Linear trend
Pacific Coast
-------- Departure from normal
——— Linear trend
Prairies
-------- Departure from normal
——— Linear trend
South British Columbia Mountains
Yukon and North British Columbia Mountains

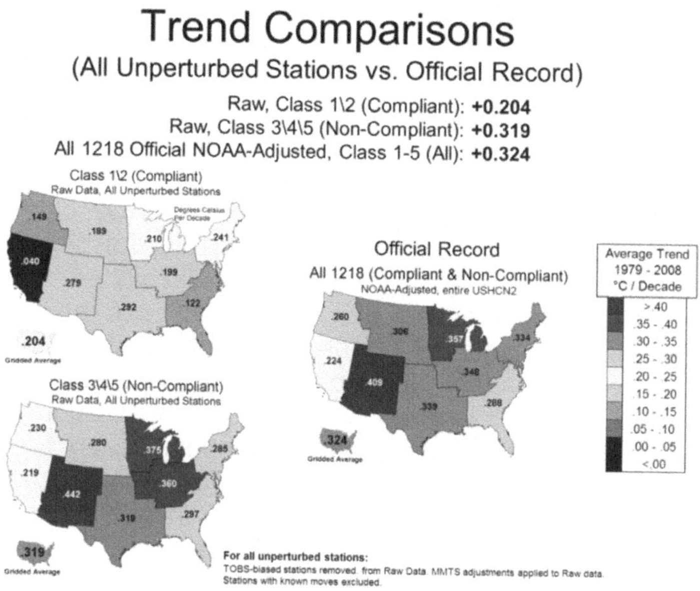


(b)

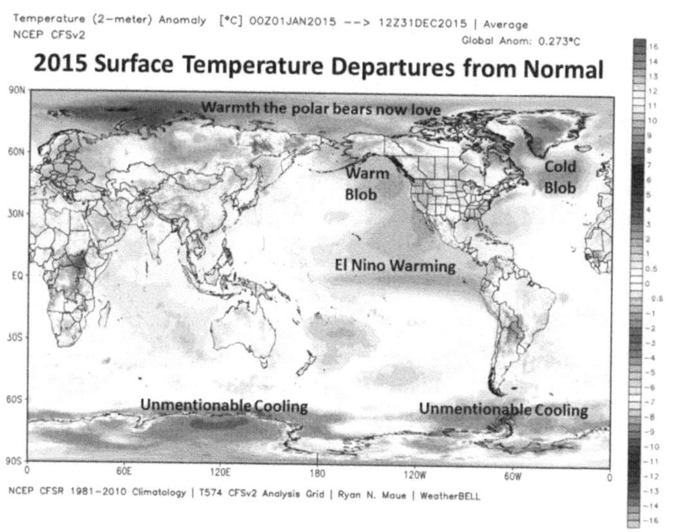


(c)

**Figure 2.1 (a)** Annual mean Temperature departures from 1961-90 normal and linear trends for different regions in Canada, 1948-2009, **(b)** The quality of temperature sitting versus temperature trend in 9 US climate zones, & **(c)** 2015 Global Surface Temperature Departures from Normal

**Figure 2.1c** presents the 2015 Global Surface Temperature Departures from Normal. The Polar Regions present exactly opposite patterns -- warmer North Pole & cooler South Pole. Such a scenario is basically results from general circulation pattern that help energy transport based on winds [direction & speed]. It is common feature that the "circumpolar Vortex" formation is linked to Stratospheric Ozone depletion. We have seen the ozone-hole formation in South Pole Region. This was explained as: in the case of South Polar Region, the strong circumpolar vortex formation limits the warm air [advection] from middle latitudes moving in to South Polar Region and thus colder conditions are strengthened with the winter six months with zero energy reaching from the Sun. The formation of vortex is more frequent in South Polar Region. In the case of North Polar Region, the circumpolar vortex is always weak and thus warm air [advection] from middle latitudes moves in to North Polar Region. This counters the six months zero energy reaching this region from the Sun. The formation of vortex is less frequent over the North Polar Region. **Figure 2.1c** presents one such case. However, the oscillations in temperature over middle latitudes generally modify the condition. Also, these temperatures are also influenced by Polar Oscillations. Unless we take all these conditions in to account, we may not reach to right conclusion.

**Figure 2.2** presents the annual march of temperature anomalies for NH, SH, Land, Ocean & Globe are presented. Ecological changes are more in NH and less in SH and zero over ocean based on the land cover. Majority of ground based data that goes in to the averaging of global average temperature is ecological changes part only as majority of the met stations are located in urban areas around the world. Because of this SH temperature pattern following ocean temperature and NH temperature pattern is following land temperature pattern [**Reddy, 2008**]. All these suggest there is no global warming associated with anthropogenic greenhouse gases concentration increase in the atmosphere as the energy versus $CO_2$ reached a saturation point with the natural $CO_2$. Also, a recent study reports that $CO_2$ is removed from the atmosphere by natural sinks at a rate proportional to the surplus $CO_2$ accumulation. The half life of the surplus $CO_2$ concentration is approximately 40 years [not 1000 years].

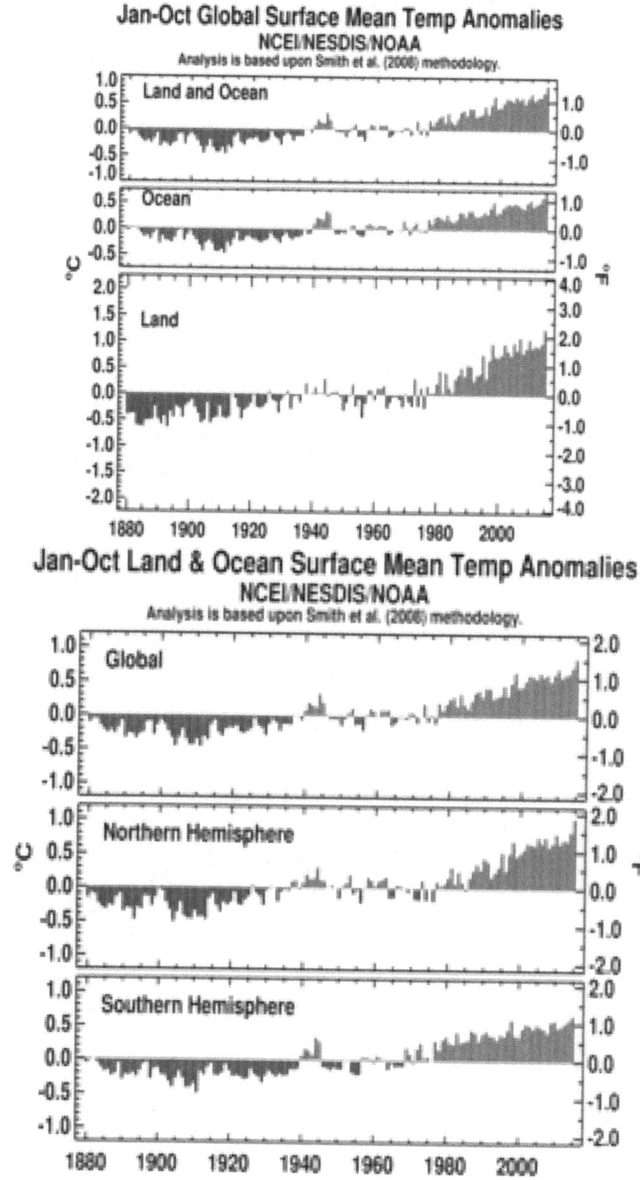


January–October Global Hemisphere plot

The most current data October be accessed via the Global Surface Temperature Anomalies page.

**Figure 2.2** Global Temperature Variation for NH & SH and Land & Ocean

**Figure 2.3** presents a classical example of data manipulation, wherein it states that "NASA cools the past". **Figure 2.4** presents the met network in Australia wherein it shows that some areas have dense network and large areas have sparse network. **Figure 2.5** presents bias of short records in data periods in Australia. These scenarios are not limited to these countries alone but can be seen all around the globe.

Every month, readings from thousands of land-based weather stations around the world are shared through the Global Historical Climatology Network. To measure ocean temperatures, there is a flow of data from buoys and ships. Climate trends play out over long periods of time, and the challenge has been to deal with changes in the way temperature is measured that have nothing to do with the weather itself. For instance, local officials might move a station from a valley to a nearby hilltop. They might change the time of day when they record their measurements from sunrise to sunset. They might change the kind of thermometer they use. In the ocean, the practice once was to haul up a bucket of water. Later, the standard practice was to measure the temperature from the engine's intake valve. All these affect the local temperature trends.

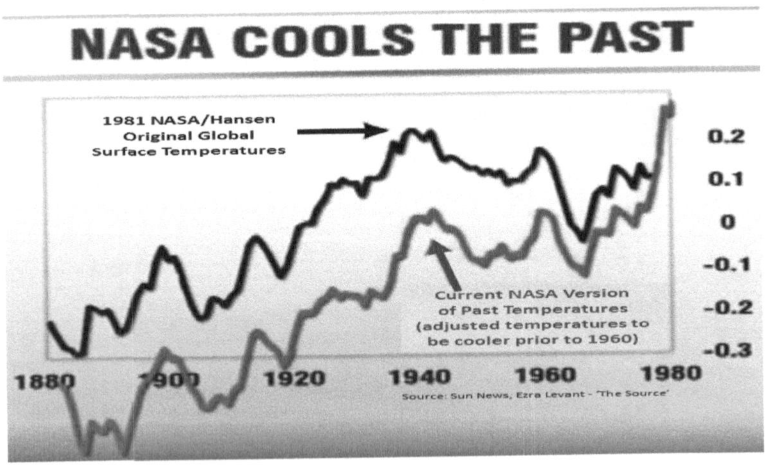


**Figure 2.3** NASA cools the past

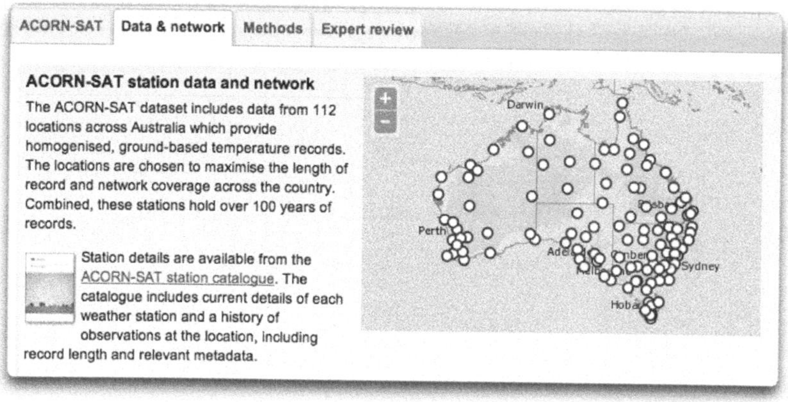


**Figure 2.4** ACORN-SAT/Australia stations network

To counter these, a report state that researchers at the NCDC/NOAA has to then come up with a way to work with the temperature readings: What they do is take the raw temperature readings and apply a boatload of statistical techniques to pick out the most reliable data. Where necessary, they adjust the readings to account for the human factors that would skew the data regardless of what happened with actual temperatures. The controversy voiced by Homewood and others is that they don't accept those adjustments. NOAA says the adjustments -- as was the case in Paraguay -- are necessary to make valid comparisons. "Such changes in observing systems cause false shifts in temperature readings. Paraguay is one example of where these false shifts artificially lower the true station temperature trend," the agency said. Homewood is right that the Paraguay adjustments raised the temperature reported for that station. But what Homewood leaves out, NOAA says, is that nearly half the time the adjustments made by researchers lower the temperature below what was actually recorded.

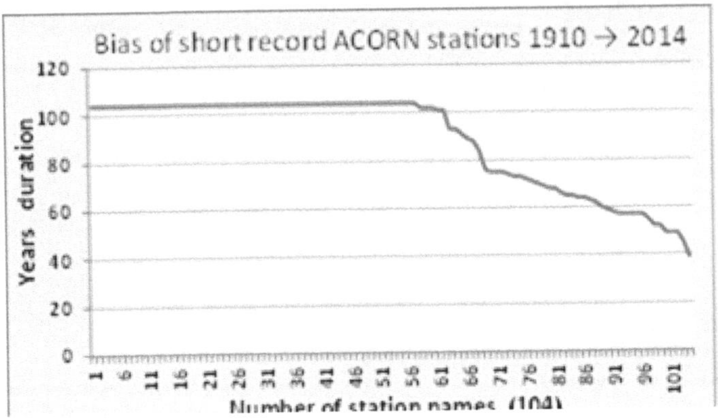


**Figure 2.5** Bias of short record ACORN/Australia stations [1910-2014]

The report further quoted some groups to support his version. This is not the reality. During the last 19 years global temperature presented hiatus – no increasing trend but the agency manipulated and showed an increasing trend in place of hiatus. – There is some controversy in using the words "pause" and "hiatus" synonymous.

(They) look through all those thousands of stations, find a few that show big adjustments, and tell everyone that they are evidence of fraud," Hausfather said. "You will rarely see them pick out stations like Reno, Paris, London, Tokyo, or many others where the adjustments dramatically lower the warming trend." Hausfather and his colleagues traced how the adjustment methods changed the temperature data differently around the world since 1850.

In the United States, with about 5 percent of Earth's land area, the official data file raised temperatures compared to the original readings. But the same methods lowered the data records in Africa and for all land-based readings taken together, the adjustments basically made no change at all. With ocean temperature trends, the efforts to compensate for the human factor lower the numbers dramatically. "The net effect of adjustments is to actually reduce the amount of global warming we've observed since 1880 by about 20 percent," Hausfather said. "Folks skeptical of temperature adjustments are welcome not to use them if they'd like, but you end up with more global warming, not less." Then the basic question is, why mutilate the data?

Mark Serreze is professor of geography at the University of Colorado-Boulder. Serreze said over the years, many people have vetted the statistical methods that lead to the adjusted data files. "The features of the global temperature records have been verified by comparisons between analyses from different centers and sensitivities to different ways of treating the data," Serreze said. "The peer reviewed literature is extensive. This is why there is consensus that the data are correct." For the record, the author of the *The Telegraph* opinion piece made the same point about eight months ago. Back then, it inspired a similar flurry of claims that government scientists intentionally engaged in fraud.

The researchers we contacted, who have no ties to the government agencies that produce the data Perino questioned, said that while the raw temperature readings are adjusted, the result is a record that more closely matches what has actually taken place. The greater mistake would be to take the raw data as somehow perfect and unblemished.

All this argument of the author of the article did prove nothing except accepting the fact the data was corrected. WMO has provided guidelines to nations on the correcting the data gaps, changes in location or changes in instruments, etc. So, it is the responsibility of individual meteorological departments and nobody put this in the hands of some data managers like NOAA or for that matter in any other agency.

## 2.2.2 Human Activity

We can categorize human activity in to three distinctive systems. They impact weather directly or indirectly. They are [a] directly impact global temperature, known as global warming, [b] directly impact weather & climate at local and or regional level, known as ecological changes, and [c] the intensity of impact is modified indirectly through physical actions. Wikipedia noted that IPCC reported in 2014 that scientists were more than 95% certain that global warming is being caused mostly by increasing concentrations of greenhouse gases and other human (anthropogenic) activities, that is [a] and [b]. However, it is not so as the former is global in nature and latter are local & regional in nature. Let us see these issues.

In a 2008 NASA article on usage, Erik M. Conway defined *Global warming* as "the increase in Earth's average surface temperature due to rising levels of greenhouse gases", while *Climate change* was "a long-term change in the Earth's climate, or of a region on Earth. "As effects such as changing patterns of rainfall and rising sea levels would probably have more impact than temperatures alone, he considered "global climate change" a more scientifically accurate term, and like the IPCC, the NASA website would emphasize this wider context. Here the author conspicuously used divergent words – a region at place and at other place used global. Changes in climate parameters is location-region specific. There is no such scenario as that of global scale change.

Pro-global warming groups argue that "Overwhelming scientific consensus says human activity is primarily responsible for global climate change". The 2010 Anderegg study found that 97-98% of climate researchers publishing most actively in their field agree that human activity is primarily responsible for "global climate change". The study also found that the expertise of researchers unconvinced of "human-caused climate change" is "substantially below" that of researchers who agree that human activity is primarily responsible for climate change. The 2013 Cook review of 11,944 peer-reviewed studies on climate change founded that only 78 studies (0.7%) explicitly rejected the position that humans are responsible for "global warming". A separate review of 13,950 peer-reviewed studies on climate change found only 24 that rejected "human-caused global warming". A survey by German Scientists Bray and Von Storch found that 83.5% of climate scientists believe human activity is causing "most of recent global climate change". A separate survey in 2011 also found that 84% of earth, space, atmospheric, oceanic, and hydrological scientists surveyed said that "human-induced global warming" is occurring.

More than one thousand scientists disagree that human activity is primarily responsible for "global climate change". In 2010 Climate Depot released a report featuring more than 1,000 scientists, several of them former UN IPCC scientists, who disagreed that humans are primarily responsible for "global climate change". The Cook review of 11,944 peer-reviewed studies found 66.4% of the studies had no stated position on "anthropogenic global warming", and while 32.6% of the studies implied or stated that "humans are contributing to climate change", only 65 papers

(0.5%) explicitly stated "that humans are the primary cause of recent global warming." A 2012 Purdue University survey found that 47% of climatologists challenge the idea that humans are primarily responsible for climate change and instead believe that climate change is caused by an equal combination of humans and the environment (37%), mostly by the environment (5%), or that there's not enough information to say (5%). In 2014 a group of 15 scientists dismissed the US National Climate Assessment as a "masterpiece of marketing," that was "grossly flawed," and called the NCA's assertion of human-caused climate change "NOT true."

In the above two Paras, either global climate change or human induced climate change or human-induced global warming or global warming or climate change or most of recent global climate change were used to refer "global warming" only. In this the main confusion is use of wording – non-uniformity. People without understanding the difference between climate change and global warming is conducting surveys, have no meaning. Climate change is not global warming. IPCC itself has agreed that the global warming is a component of global temperature raise through greenhouse effect due to increased levels of greenhouse gases released by fossil fuel use after industrialization, a human activity. It is presumed as global component.

There is another human activity associated with changes in ecology – land & water use and land & water cover. It is a localized and or regionalized component unlike the global warming. Though it goes in to global temperature averaging, it has no influence on nature at global scale. This influences the weather & climate at local and regional scales. The systematic variations component of climate change, which is a local and regional phenomenon, is influenced by the ecological changes component. The global average temperature raise is biased by this component in the overestimation. The other important factors relate to aerosols and filth cover that are going to influence the temperature.

## 2.2.3 Trends & Anomalies

In nature, the greenhouse gases effect follow inverted 'Z' shape pattern in conversion of energy in to temperature with the increase of greenhouse gases in the atmosphere **[Reddy, 1995a]** as the energy factor is not unlimited but it has a "cap". Ken Stewart on 16th December 2015 presented an article "Energy, Carbon Dioxide, and The Pause" by using energy data from the BP Statistical Review of World Energy 2015, $CO_2$ data from NOAA, and Temperature data from UAH. Between 336 to 400 ppm of $CO_2$ presented a relation of 0.0063 X, and between 370 to 400 ppm of $CO_2$ presented 0.0004 X [pause]. This is exactly what I presented in an article with reference to relative growth or relative yield versus relative radiation stress or relative water stress or relative nutrient stress wherein maximum impact is in the slant line part shown in inverted Z [**Reddy, 1995a**]. The horizontal portion [pause] in the present case started around 370 ppm. The initial horizontal part might have ended at around 150 ppm. The slant portion reached maximum in between 150 to 370 ppm of $CO_2$. With more accurate temperature data that takes in to account the rural-cold-island effect in addition to accuracy factor will provide the correct limits to $CO_2$ for starting of slant portion and starting of pause.

Some argued that the term "Global" warming, no one can relate to such a temperature and it does not do justice to what is really happening globally. Here is an example of January to December land temperature anomalies trend for the last 10 years, namely 2005 – 2014. The data is as per NOAA climate at a glance [units in °C/decade]: *Contiguous US (-0.68); North America (-0.41); Asia (-0.31); Europe (+0.39); Northern Hemisphere (-0.04) flat; Oceania (+0.07) flat; Africa (+0.08) flat; South America (+0.24); Southern Hemisphere (+0.06) flat; Global (-0.01) flat.*

Canada has not shown global warming. **Figure 2.1a** presents the annual mean temperature departures from 1961 to 1990 normal over different regions in Canada for the period 1948 to2009 [www.ec.gc.ca/adsc-cmda/default.asp?lang=En&n=F3D25729-1]. The trends show high variations over different parts of Canada. Some areas like Northeastern Forests, Atlantic Canada, Greater lakes and St. Lawrence area show low variations; and some other regions show very high variations. Around the last decade there are sudden jumps that raise the trend significantly also inter-annual variability show very high in last half period. All these clearly show that this is nothing to do with global warming but it is associated with the ecological changes – land use & land cover and water use & water cover changes [Reddy, 2008].

**Table 2.1** presents the ranges and average anomaly estimated based on 37 years [during December 1978 to November 2015] of

NOAA/NASA Satellite data. This is something fishy – manipulated data – that [1] all the land was warming at the same level except that of South Pole –– 0.18 to 0.21 °C and [2] all the ocean except at poles –– 0.07 to 0.09 °C.

**Table 2.1** Global satellite temperature data trends and average anomalies under different groups

| **Trend and average anomaly in 37 year satellite temperature data, °C** | | | | | |
|---|---|---|---|---|---|
| **Area** | **Trend/decade** | **Anomaly** | **Area** | **trend/decade** | **Anomaly** |
| Globe | +0.11 | +0.41 | Tropics | +0.10 | +0.37 |
| Land | +0.20 | +0.74 | Land | +0.21 | +0.38 |
| Ocean | +0.08 | +0.30 | Ocean | +0.07 | +0.26 |
| NH | +0.14 | +0.52 | N. Pole | +0.23 | +0.8 |
| Land | +0.21 | +0.78 | Land | +0.19 | +0.70 |
| Ocean | +0.09 | +0.33 | Ocean | +0.26 | +0.96 |
| SH | +0.19 | +0.33 | S. Pole | -0.01 | -0.04 |
| Land | +0.18 | +0.67 | Land | -0.01 | -0.04 |
| Ocean | +0.07 | +0.26 | Ocean | -0.01 | -0.04 |
| USA48 | +0.19 | +0.70 | Australia | +0.24 | +0.89 |

From all these it is clear that, the warming is not really warming at global scale but it is sum and total of what is happening at local and regional levels only. The alarmists claim global warming is happening at an unprecedented rate. Yet it is only happening for the oceans and that to only due to data manipulation. Where people live it is an entirely different picture. Some areas are cooling some are warming and some areas are not changing at all or flat. So a single "global" warming figure is a misrepresentation of what is really happening around the world.

Heating and cooling is part of natural systems. Also, greenhouse effect is also not new. They are there for centuries. The question to be answered is: How much the two man-induced components in temperature are influencing them?

Wikipedia writes that the possible societal responses to global warming include mitigation by emissions reduction, adaptation to its effects, building system resilient to its effects, and possible future climate engineering. Here we must remember the fact that during the doubling period all human activities, both [a] & [b] were more than doubled. That is not only greenhouse gases raise but also ecological changes. Though for cutting emissions that future global warming should be limited to below 2.0 °C relative to pre-industrial level but the actual data refers to the global average (land and water) surface temperature which includes other than the emission induced temperature rise as mentioned. Unless this part is not known quantitatively, how it is possible to specify the quantum of reduction in emissions and thus when the 2 °C will reach under that reduction scenario: in 100 years, or 1000 years, or 10000 years, etc.

Temperature changes vary over the globe. Since 1979, land temperatures have increased about twice as fast as ocean temperatures (0.25 °C per decade against 0.13 °C per decade; which means, average of 1.7 °C per Century). Ocean temperatures increase more slowly than land temperatures because of the larger effective heat capacity of the oceans and because the ocean loses more heat by evaporation. Since the beginning of industrialization the temperature difference between the hemispheres has increased due to melting of sea ice and snow in the North. Average Arctic temperatures have been increasing at almost twice the rate of the rest of the world in the past 100 years; however Arctic temperatures are also highly variable. Although more greenhouse gases are emitted in the Northern than in the Southern Hemisphere this does not contribute to the difference in warming because the major greenhouse gases persist long enough to mix between hemispheres. Because of this, this is a global phenomenon but other human activity related raise in temperature is not global but it is local and regional. There is vast difference in temperature trend in the Southern Hemisphere [SH] and the Northern Hemisphere [NH] as in the former more area is covered by the Oceans and the later by the Land and thus ecological changes contributed to higher temperatures in NH and less to SH. So, it is not [a] but it is [b] that is contributing to global temperature rise. [a] refers to global warming. This clearly reflects the importance of ecological changes component which has large Hemispherical differences [**Reddy, 2008**].

Fourteen of the fifteen hottest years have been in the 21st century. While record-breaking years can attract considerable public interest, individual years are less significant than the overall trend. So some climatologists have criticized the attention that the popular press gives to "warmest year" statistics; for example, Gavin Schmidt stated "the long-term trends or the expected sequence of records are far more important than whether any single year is a record or not. This is true, as they are part of systematic variations in the climate change. Also, it is quite obvious as the contribution of ecological changes is going on added.

Statistical uncertainty means that it is hard to separate years where the temperatures are close, so while 2014 had a greater chance than any other single year, it only has a 38% to 48% likelihood of being the warmest year on record since 1880. 2014 was the 38th consecutive year with above-average temperatures. Ocean oscillations like El Nino [Southern Oscillation] (ENSO) can affect global average temperatures, for example, 1998 temperatures were significantly enhanced by strong El Niño conditions. 1998 remained the warmest year until 2005 and 2010 and the temperature of both of these years was enhanced by El Niño periods. However, 2014 was ENSO neutral. Here we must keep in mind is that on either side of the peak drops are seen.

If we look in to the NOAA update "ENSO: Recent Evolution, Current Status and Prediction", 28th December 2015, in the last one year there are clear cut zones with positive and negative temperature anomalies both at surface and deeper layers of the oceans. For example, during January to Mid-March 2015, near-to-below average SSTs were observed in the eastern Pacific, and positive SST anomalies persisted across the western and central Pacific; from June to September, the largest positive SST anomalies shifted westward; and recently, positive SST anomalies greater than 2 $^{o}$C extended from 170 $^{o}$W eastward to the coast of South America. We mostly talk of positive rise in temperature but we rarely look in to negative fall in temperature. We rarely look in to shift of positive and negative zones with the passing of time. All these have significant influence on the weather systems existing in those respective zones. So, average has no meaning yet all to understand the weather.

The other important issue is the accelerated increase in SSTs in recent years. Whether the average global temperature or the average SSTs they are influenced by cyclic variation – 60-year cycle. When you present a trend for different segments of the data series from 1880 to date, you get different trends. But, if the interest is to understand the realistic trend [whether it is accelerating or decelerating] first we must understand, whether the data is following a cyclic variation; if it is following then remove that cyclic part using moving average. Then you get the trend. Here you can try to divide the data in to segments based on trend. We rarely find such analysis as the intension is not science.

**Table 2.2** Global, land & ocean temperature anomalies for 1880-2014 & 1951-2012 with old & new data series

| Global, land & ocean temperature anomalies, °C/decade | | | | |
|---|---|---|---|---|
| **Period** | **Data type** | **Global** | **Land** | **Ocean** |
| 1880-2014 | Old | 0.068±0.017 | 0.106±0.017 | 0.055±0.017 |
| | New | 0.065±0.015 | 0.096±0.016 | 0.055±0.0.15 |
| 1951-2012 | Old | 0.129±0.020 | 0.207±0.031 | 0.100±0.017 |
| | New | 0.117±0.021 | 0.194±0.031 | 0.088±0.017 |

[Source: Karl, et al, 2015]

We are dealing with manipulated data series. **Table 2.2** presents the NOAA's New "Pause-Buster" SST data – Karl, et al, 2015 – for 1880 to 2014 and 1951 to 2012 averages of global, land & ocean temperature anomalies with old and new data series. Practically there is no difference between new and old temperature anomalies during 1880-2015 in the case of global & oceans; and also, global values are very close to ocean values. Land value is nearly double to that of ocean value. In the case of 1951-2012 data series, in all the three the difference between old and new is around 0.012 to 0.013. It is quite conspicuous the fact that the land value is double to the ocean value in both the series; and the error variations are the same except for land under the 1951-2012 data series. This show there is something fishy in the data. Why I am saying this is, basically because the non-greenhouse effect component is insignificant over oceans while in the land value its contribution is significantly large. Also, the filth reduces the energy reaching the ocean waters and we get reduced temperature. Therefore while the land value increases, ocean value decreases.

Let me present a case study relating the precipitation data of state of Andhra Pradesh in India. This region receives rainfall in two monsoon seasons, namely southwest monsoon & northeast monsoon. In addition it receives rainfall in association with the cyclonic activity in pre-monsoon and post monsoon [northeast monsoon] seasons. The northeast and southwest monsoon precipitation showed a 56 year cycle but in opposite direction.

Here the cyclonic activity followed the northeast monsoon 56-year cyclic pattern only. Both the monsoon precipitation between 1871 and 1994 presented an increasing trend [**Figure 3.16**]. However, the annual data that takes in to account all the precipitation giving systems presented a 132 year cycle [**Figure 3.18**], wherein the first 66 year followed below the average pattern with 24 drought years and 12 flood years. The next 66 years followed the above the average pattern with 24 flood years and 12 drought years. The present 132 year cycle started in 2001 wherein majority of the years presented below the average or drought years. Therefore, unless make effort to understand the variations in wider spectrum, we go in to false syndrome.

Weather aberrations are attributed to global warming phenomena. For example, to show hiatus or pause in global average temperature anomaly stopped warmists groups bombarded the media with slogans "Warmest 2015". Here the weather aberration [not a part of trend] associated with the El Nino showed a rise in temperature. Generally, on either side of El Nino peak present depressions. The current El Nino depression will appear by 2016/17. 2015 temperature anomaly showed excess by 0.13, 0.18 or 0.16 °C depending upon the choice of the data set with ± 0.10, 0.10 or 0.09 °C. That means they vary between 0.003 to 0.23, 0.08 to 0.28 or 0.06 to 0.26 °C. Temperature is reported to the 1st place of decimal only. Then they are given as 0.0 to 0.2, 0.1 to 0.3 or 0.1 to 0.3 °C from 2014. The rise with El Nino was expected by 3.0 °C. So, 2015 is in hiatus path only.

## 2.2.4 History of Temperature

A 2003 study by researchers at the Harvard-Smithsonian Center for Astrophysics found that "many records reveal that the 20th century is probably not the warmest or a uniquely extreme climatic period of the last millennium "A 2005 study published in *Nature* found that "high temperatures - similar to those observed in the twentieth century before 1990 - occurred around AD 1000 to 1100" in the Northern Hemisphere. A 2013 study published in *Boreas* found that summer temperatures during the Roman Empire and Medieval periods were "consistently higher" than temperatures during the 20th century. According to a 2010 study in the *Chinese Science Bulletin*, the recent global warming period of the 20th century is the result of a natural 21-year temperature oscillation, and will give way to a "new cool period in the 2030s."

Average temperatures on the Earth have increased at a rate far faster than can be explained by natural climate changes. A 2008 study compared data from tree rings, ice cores, and corals over the past millennium with recent temperature records. The study created the famous "hockey stick" graph, showing that the rise in earth's temperature over the preceding decade had occurred at a rate faster than any warming period over the last 1,700 years. In 2012 the Berkeley scientists found that the average temperature of the earth's land increased 2.5°F over 250 years (1750-2000), with 1.5°F of that increase in the last 50 years. Lead researcher Richard A. Muller, PhD, said "it appears likely that essentially all of this increase [in temperature] results from the human emission of greenhouse gases." In 2013, a surface temperature study published in *Science* found that global warming over the past 100 years has proceeded at a rate faster than at any time in the past 11,300 years. According to the IPCC's 2014 Synthesis Report, human actions are "extremely likely" (95-100% confidence) to have been the main cause of 20th century global warming, and the surface temperature warming since the 1950s is "unprecedented over decades to millennia."

Natural changes in the sun's activity cannot explain 20th century global warming. According to a Dec. 2013 study in *Nature Geosciences*, the sun has had only a "minor effect" on the Northern Hemisphere climate over the past 1,000 years, and global warming from human-produced greenhouse gases has been the primary cause of climate change since 1900. Another 2013 study found that solar activity could not have contributed to more than 10% of the observed global warming over the 20th century. Measurements in the upper atmosphere from 1979-2009 show the sun's energy has gone up and down in cycles, with no net increase. According to a 2013 IPCC report, there is "high confidence" (8 out of 10 chance) that changes in the sun's radiation could not have caused the increase in the earth's surface temperature from 1986-2008. Although warming is occurring in the lower atmosphere (troposphere), the upper atmosphere (stratosphere) is actually cooling. If the sun were driving global warming, there would be warming in the stratosphere also, not cooling.

Global warming and cooling are primarily caused by fluctuations in the sun's heat (solar forcing), not by human activity. Over the past 10,000 years, solar minima (reduced sun spot activity) have been "accompanied by sharp climate changes." Between 1900 and 2000 solar irradiance increased 0.19%, and correlated with the rise in US surface temperatures over the 20th century. According to a 2007 study published in *Energy & Environment*, "variations in solar activity and not the burning of fossil fuels are the direct cause of the observed multiyear variations in climatic responses." In a 2012 study by Willie Soon, PhD, Physicist at the Harvard-Smithsonian Center for Astrophysics, a strong correlation between solar radiation and temperatures in the Arctic over the past 130 years was identified. According to a 2012 study published in the *Journal of Atmospheric and Solar-Terrestrial Physics*, "up to 70% of the observed post-1850 climate change and warming could be associated to multiple solar cycles."

The rate of global warming has slowed over the last decade even though atmospheric $CO_2$ continues to increase. IPCC recognized a slowdown in global warming over the past 15 years in its 2013 report. According to the Heartland Institute's 2013 NIPCC report, the earth "has not warmed significantly for the past 16 years despite an 8% increase in atmospheric $CO_2$." In Aug. 2014 a study in the *Open Journal of Statistics* analyzed surface temperature records and satellite measurements of the lower atmosphere and confirmed that this slowdown in global warming has occurred. According to Emeritus Professor of Meteorology at the Massachusetts Institute of Technology Richard Lindzen, Ph.D., "the IPCC's excuse for the absence of warming over the past 17 years is that the heat is hiding in the deep ocean. However, this is simply an admission that the [climate] models fail to simulate the exchanges of heat between the surface layers and the deeper oceans..."

Predictions of accelerating human-caused climate change are based upon computerized climate models that are inadequate and incorrect. Climate models have been unable to simulate major known features of past climate such as the ice ages or the very warm climates of the Miocene, Eocene, and Cretaceous periods. If models cannot replicate past climate changes they should not be trusted to predict future climate changes. A 2011 *Asia-Pacific Journal of Atmospheric Science* study using observational data rather than computer climate models concluded that "the models are exaggerating climate sensitivity" and overestimate how fast the earth will warm as $CO_2$ levels increase. Two other studies using observational data found that IPCC projections of future global warming are too high. In a 2014 article, climatologist and former NASA scientist Roy Spencer, Ph.D., concluded that 95% of climate models have "over-forecast the warming trend since 1979." According to Emeritus Professor of Geography at the University of Winnipeg, Tim Ball, Ph.D., "IPCC computer climate models are the vehicles of deception... [They create the results they are designed to produce."

## 2.3 Global Warming versus its Impacts

### 2.3.1 General

Tom Harris presented an article "Nine Experts slam EPA Administrator Gina McCathey's Power Plan" in PJMedia on August 24, 2015. I am one among those nine experts. In McCathey presentation at the Resources for the Future Policy Leadership Forum, her first public appearance since August 3 release of the EPA's "Clean Power Plan" of USA. Let us see some of the observations made by McCathey in her speech and response by the nine experts as well as Tom Harris, the author of the article.

McCathey observed "Climate change is one of the most important issues that we face. It is a global challenge but, in many ways, it's also very personal to all of us because it affects everything and everyone we know and we love." Climate change is, of course, a regional challenge, not a global one [*This exactly what I was saying for the last three decades*]. There is no super being straddling the planet, experiencing global trends. All that matters is what is happening increases or decreases in the incidence of floods and droughts, heat waves and cold spells, and so on in regions where people, plants, and animals are found. For example, what sense would it make for a community to prepare for a global sea level rise if, in that particular region, sea level was falling? Bryan Leyland countered "Climate change has been a problem to mankind for hundreds of thousands of years. But we survived the last ice age, compared with which, the recent change in climate is but a minor wiggle. The greatest climate risk we face at the moment is a high probability that we are entering a period of cooling comparable to the Little Ice Age." Many scientists agree with Leyland. For example, Dr. Howard Hayden emeritus professor of physics at the University of Connecticut explained: "The Earth is on a descent into the next 100,000-year ice age. For the moment, the glaciers seem to be in retreat, but they are *not* remnants of the last ice age. They have been growing during the last 8,000 years." Dr. John Nicol, former senior lecturer of physics and dean of science at James Cook University in Australia, elaborated: "Since 1997, the Earth has not warmed but has, in fact, very slightly cooled even though atmospheric $CO_2$ levels have been increasing. McCarthy's assertion that climate change is "very personal to all of us" clearly demonstrates her emotional rather than the scientific approach to this non-issue."

Gina McCarthy next said: "By now we all know that climate change is driven in large part by carbon pollution and it leads to more extreme heat, cold, storms, fires, and floods." Leyland replied to the EPA chief: "It is shocking that McCarthy does not understand the difference between carbon dioxide a harmless gas that benefits agriculture and genuine pollutants like particulates, sulphur dioxide and the like emitted from old obsolete power stations. Modern coal-fired stations do not emit these pollutants." Regardless, neither theory nor observations support the EPA chief's claim that $CO_2$ rise causes "more extreme heat, cold, storms, fires and floods." Hyderabad, India-based Dr. S. Jeevananda Reddy formerly chief technical advisor for the UN World Meteorological Organization and author of *Climate Change – Myths & Realities* said that McCarthy's statement is "not true": "Extreme heat, cold, storms and floods are part of natural variation. These are modified by local general circulation patterns existing over different parts of the globe over different seasons." Nicol also contested McCarthy's assertion: "Not only is the claim that $CO_2$ is to blame [for increases in extreme weather] wrong, but the contradictory statements regarding these weather events, which are NO different from those of 200 years ago, demonstrates the desperation of lobby groups trying to maintain this myth. If the world were to warm appreciably due to increasing $CO_2$ emissions, temperatures at high latitudes are forecast to raise the most, reducing the difference between arctic and tropical temperatures. Since this differential drives weather, we should see weaker mid-latitude cyclones in a warmer world and thus fewer extremes in weather, not more."

Tom Harris noted "Indeed, the lack of extreme weather increase with global warming is one of the few areas of agreement between the United Nations Intergovernmental Panel on Climate

Change (IPCC) and the Nongovernmental International Panel on Climate Change (NIPCC). In 2012, the IPCC asserted that a relationship between global warming and wildfires, rainfall, storms, hurricanes, and other extreme weather events has not been demonstrated. In their latest assessment report (Sep 2013), IPCC scientists concluded that they had only "low confidence" that "damaging increases will occur in either drought or tropical cyclone activity" as a result of global warming. The Sep 2013 NIPCC report concluded the same, asserting: "In no case has a convincing relationship been established between warming over the past 100 years and increases in any of these extreme events. NIPCC report chapter lead author Dr. Timothy Ball, environmental consultant and former climatology professor at the University of Winnipeg, explained that the EPA is taking the approach that American Journalist Farhad Manjoo identified in his book *True Enough: Learning To Live in a Post-Fact Society*: "You create your theory then hire experts. The EPA agenda is political, not scientific."

McCarthy then told the RFF Forum: "For farmers who are strained by the drought, for families with homes in the path of a wildfire, for small businesses along our coastlines, climate change is indeed very personal." Nicol writing: "Farmers do not believe in Global Warming or Climate Change as spruced by the human-caused global warming industry. Farmers have mostly been on their properties since they were children and have also been given detailed accounts of the weather and the seasons from when their great-grandfathers began farming. This fact upsets those who try to claim that there are obvious changes. Farmers will tell you that the seasons come in cycles and any season we have now has been seen in the past — possibly 100 years ago." Reddy also replied to McCarthy: "These [phenomena McCarthy lists] are associated with human actions on nature — land use and land cover changes, pollution (air, water, soil, and food) and adulterated foods, etc. For example, recent devastations in Jammu & Kashmir and Himalayan states of India were associated with occupation/building houses on river banks."

McCarthy continued: "We all know that climate change is impacting us today and will continue to get worse if we don't take action." The EPA chief knows full well that this is not true. After intense questioning from Representative Mike Pompeo (R-KS) at the September 18, 2013 hearing of the House Subcommittee on

Energy and Power, McCarthy admitted that the CPP will have essentially no impact on climate. Hayden agreed: "Even if the restrictions were enacted, the effect on worldwide temperature would be too tiny to measure."

McCarthy said: "We are way past any further discussion or debate. Scientists are as sure that humans are causing climate change as they are that cigarette smoke causes lung cancer. So, unless you want to debate that point, don't debate about climate change any longer because it is our moral responsibility to act." Tom Harris noted that "Comparing the science linking cigarette smoke and cancer with the science of climate change is ridiculous. Climate science is becoming more uncertain as the field advances — we don't even know if warming or cooling lies ahead." University of Western Ontario applied mathematician Dr. Chris Essex, an expert in the mathematical models that are the basis of the climate scare, explained: "Climate is one of the most challenging open problems in modern science. Some knowledgeable scientists believe that the climate problem can never be solved." Tom Harris noted "The NIPCC reports list hundreds of peer-reviewed science papers that show that much of what we thought we knew about climate is wrong or highly debatable. In particular, the lack of global warming over the past 18 years, a period during which $CO_2$ concentration in the atmosphere has risen 10%, shows there is something seriously wrong with the human-caused warming theory." Reddy responded to McCarthy's statement: "We still need to discuss global warming science since the IPCC is not sure of the correct sensitivity factor that relates anthropogenic greenhouse gas increases to temperature rise. This is illustrated by the fact that they changed the sensitivity factor [the temperature rise in degrees Celsius forecast to occur due to a doubling of $CO_2$], from 1.95 in the IPCC Fourth Assessment Report (2007) to 1.55 in their Fifth Assessment Report (2013). They are merely employing trial and error, and not physical process paths." Ball points out what the IPCC itself admitted in its Third Assessment Report (2001): "In climate research and modeling, we should recognize that we are dealing with a coupled non-linear chaotic system, and therefore that the long-term prediction of future climate states is not possible."

McCarthy next told the audience that Obama: "... reminded us that, while we are the first generation to feel the impacts of climate change, we are the last that can effectively do something about it."

Nicol answered: “We are not the first people to experience climate change. The Navajo in America, civilizations in the Middle East, and many others moved across continents to escape climate change-related events which were totally the responsibility of Nature and caused huge upheaval. The changes claimed to be perceived today are, by comparison, trivial. Carbon dioxide is not causing changes to the climate — Nature causes changes and always has, always will.” Ball asked: “How on Earth did we ever survive the climate change that has gone on for five billion years?”

McCarthy concluded her comments: “Science has spoken on this. A low-carbon future is inevitable. We're sending exactly the right signals on what, at least EPA believes to be, a future of lower pollution that is essential for public health and the environment.” Nicol replied: “Advocates for the destruction of society and world control of our societies are the actual offenders who have spoken on this. Real and demonstrable science shows that a low carbon future will have no influence on the world's climate and will destroy our ability to care for the world's poor. Energy is essential for the distribution of health and wealth to the poorer nations. This means that coal-fired power is essential, as recognized by the world's largest economies, China and India. Who are we to dictate the living standards of these and other nations?” Leyland added: “The main effect of the drive for a low carbon future is that energy will become more and more expensive and more and more people will die in the winter from the cold and in summer because they cannot afford to run the air conditioning. The health effects would be seriously negative. The environmental effects will be a reduction in plant growth that could cost the agricultural economy trillions of dollars.”

McCarthy made numerous excited claims about the health impacts of the new climate rule: “As a result [of the CPP], in 2030, we are going to be avoiding thousands of premature deaths and hospital admissions, tens of thousands of asthma attacks and hundreds of thousands of missed school days and missed work days.” Tom Harris observed “But the CPP does not regulate pollution. It regulates $CO_2$, which has no detrimental impact on human health. Only by assuming that enabling the CPP will force the closure of coal-fired electricity stations –and that will reduce pollution emissions – can one claim the health benefits claimed by McCarthy. As explained by William Yeatman, environmental policy expert and editor of the Cooler Heads Coalition: “[This is] an EPA scam, known as "co-benefits," by which the agency has justified a number of recent highly politicized regulations. [T]here are entire sections of the Clean Air Act given to the regulation of particulate matter and nitrogen oxides. There is, therefore, neither a public health purpose nor a need for EPA to use a climate plan to regulate particulate matter and nitrogen oxides emissions under the Clean Air Act. Furthermore, Yeatman demonstrates that the EPA's methodology for estimating health benefits of the Clean Air Act "is based almost entirely on controversial, 'secret' science." Not only do their forecasts of lives saved make no sense, but the agency refuses to release the data used to make these calculations. Carlin labeled the supposed health benefits of the CPP "dubious if not imaginary," and asked: "If these benefits actually exist, why EPA has not already obtained them directly and more efficiently using "conventional" pollutant regulations?"

Pope Francis's Christmas address [25-12-2015] which traditionally touches on the gravest issues facing the world – conspicuously by its absence was any reference to climate change or environmental issues [few months back released a document emphasizing these issues], despite the fact that less than a month ago the pope said that humanity was "on the brink of suicide" because of global warming and that the COP21 climate meetings in Paris [Nov. 30 to Dec. 12, 2015] might be the last chance for mankind to avert environmental destruction. However, the climate issue continues to divide the Catholic Church at the highest levels.

## 2.3.2 Sea Level Changes

### 2.3.2.1 Arctic & Antarctic Sea Ice Melt

The Antarctic is in some ways the precise opposite of the Arctic. This is also reflected in temperature pattern also [**Figure 2.6a & b**]. The Arctic is an ocean basin surrounded by land, which means that the sea ice is corralled in the coldest, darkest part of the Northern Hemisphere. The Antarctic is land surrounded by ocean. Whereas Northern Hemisphere sea ice can extend to roughly 40 degrees north, Southern Hemisphere sea ice can extend to roughly 50 degrees south. Moreover, Antarctic sea ice does not extend southward to the pole; it can only fringe the continent. Total Antarctic sea ice peaks in September the end of Southern Hemisphere winter historically rising to an extent of roughly 18 million square kilometers. Ice extent reaches its minimum in

February, when it dips to roughly 3 million square kilometers. Roughly 15 million square kilometers of ice alternately melt and freeze during the annual cycle. Figures 2.7 and 2.8 gives some insight to these issues of Arctic zone and Figures 2.9 and 2.10 for Antarctica zone. According to Denmark Ocean and Ice services "Arctic ice is increasing, with the extent of ice at the highest it has been since 2004". However, the 2012 scenario is only repetition of the scenario in 1889, 150 years back at Arctic.

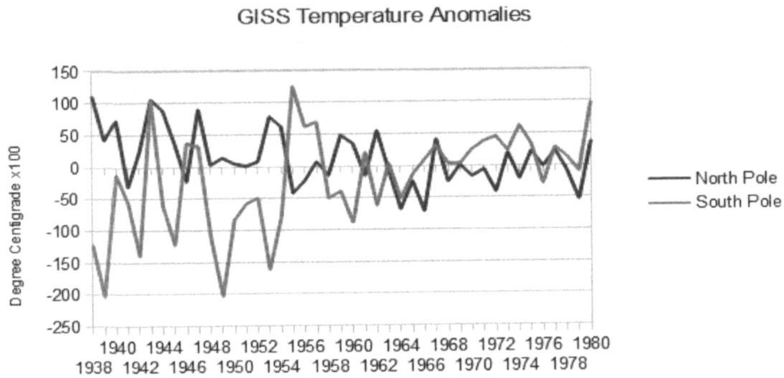


(a) 1938 to 1980

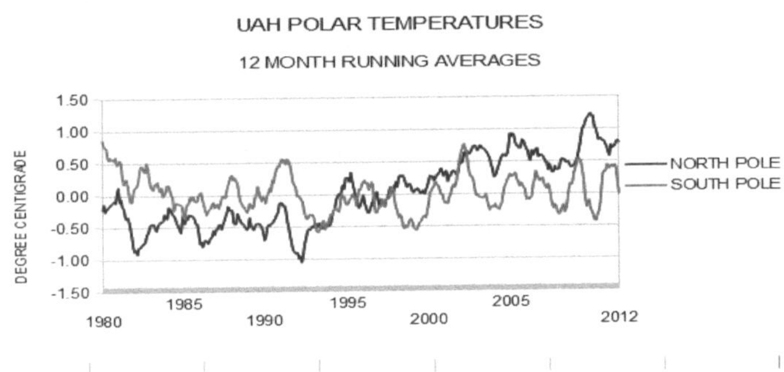


(b) 1980 to 2012

**Figure 2.6** Natural Variations in Polar Temperatures of NH & SH – [a] 1938 to 1980 & [b] 1980 to 2012

**Natural Variability:** Antarctic sea ice is distributed around the entire fringe of the continent a much broader area than the Arctic and it is exposed to a broader range of land, ocean, and atmospheric influences. Because of the geographic and climatic diversity, Antarctic sea ice is more variable from year to year than Arctic sea ice. In addition, climate oscillations don't affect ice in all sectors the same way, so it is more difficult to generalize the influence of climate patterns to the entire Southern Hemisphere ice pack.

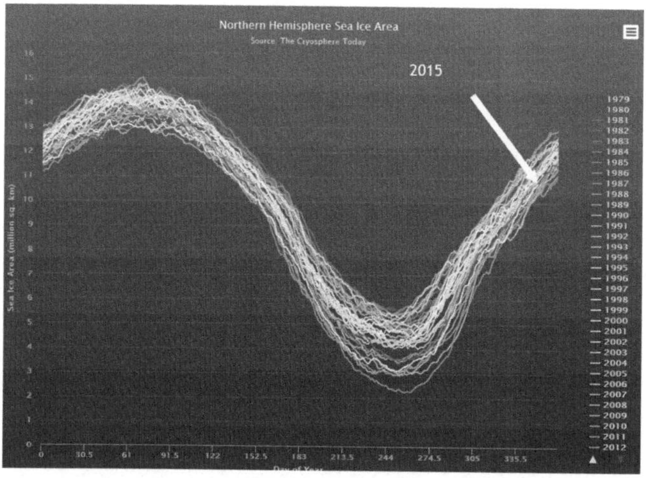


**Figure 2.7** Arctic Ice Volume: 1979-2015

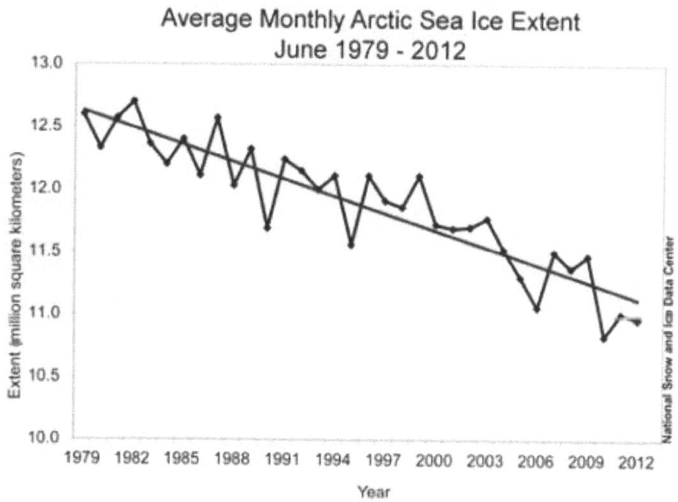


**Note**: Arctic sea ice coverage in June - the sunniest month of the year - has dropped by roughly 10 percent since 1980, far less than the drop in September coverage [Image courtesy of NSIDC].

**Figure 2.8** Average monthly Arctic Sea Ice Extent – June 1979 to 2012

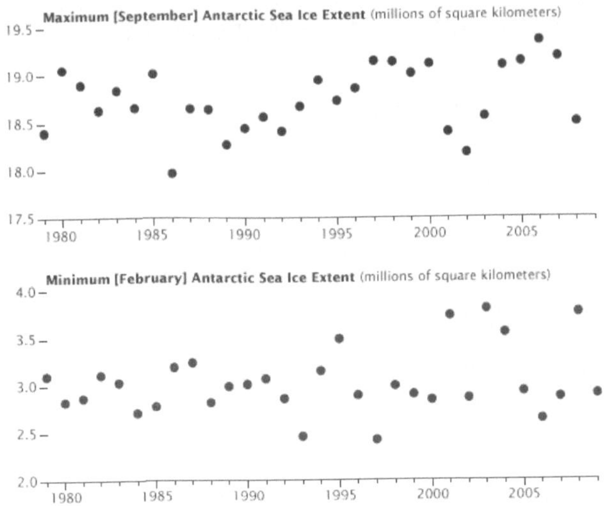


Antarctic sea ice does not plainly show the effects of global warming. There is little evidence of long-term change in either the maximum (September) or minimum (February) ice extent. (Graph by Robert Simmon, based on data from the National Snow and Ice Data Center.)

**Figure 2.9** Maximum and minimum Antarctic Sea Ice Extent, 1979 – 2008

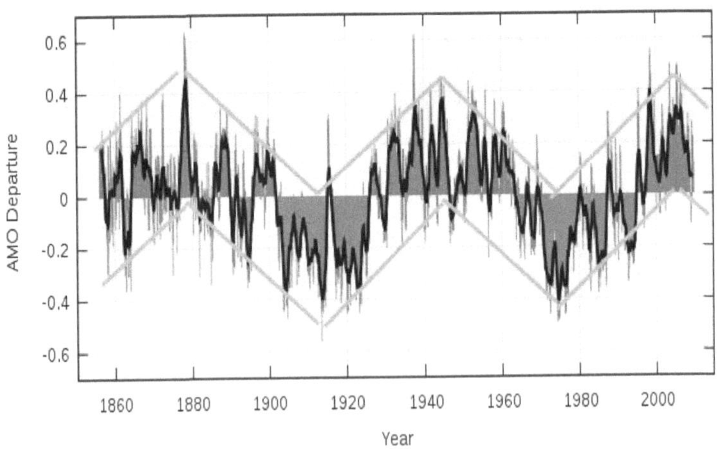


**Figure 2.10** Monthly values of the AMO Index, 1856-2009

Similar to the Arctic, the Antarctic experiences atmospheric oscillations and recurring weather patterns that influence sea ice extent. The primary variation in atmospheric circulation in the Antarctic is the Antarctic Oscillation, also called the Southern Annular Mode. Like the Arctic Oscillation, the Antarctic Oscillation involves a large-scale seesawing of atmospheric mass between the pole and the mid-latitudes. This oscillation can intensify, weaken, or shift the location of semi-permanent low- and high-pressure weather systems. These changes influence wind speeds, temperature, and the track that storms follow, any of which may influence sea ice extent.

For example, during positive phases of the Antarctic Oscillation, the prevailing westerly winds that circle Antarctica strengthen and move southward. The change in winds can change the way ice is distributed among the various sectors. In addition, the strengthening of the westerly winds isolate much of the continent and tends to have an overall cooling effect, but it causes dramatic warming on the Antarctic Peninsula, as warmer air from over the oceans to the north is drawn southward. The winds may drive the ice away from the coast in some areas and toward the coast in others. Thus, the same climate influence may lessen sea ice in some sectors and increase it in others.

Changes in the El Niño-Southern Oscillation Index (ENSO), an oscillation of ocean temperatures and surface air pressure in the tropical Pacific, can lead to a delayed response (three to four seasons later) in Antarctic sea ice extent. In general, El Niño leads to more ice in the Weddell Sea and less ice on the other side of the Antarctic Peninsula, while La Niña causes the opposite conditions.

Another atmospheric pattern of natural variability that appears to influence Antarctic sea ice is the periodic strengthening and weakening of something that meteorologists call “zonal wave three,” or ZW3. This pattern alternately strengthens winds that blow cold air away from Antarctica (toward the equator) and winds that bring warmer air from lower latitudes toward Antarctica. When southerly winds intensify, more cold air is pushed to lower latitudes, and sea ice tends to increase. The effect is most apparent in the Ross and Weddell Seas and near the Amery Ice Shelf.

As in the Arctic, the interaction of natural cycles is complex, and researchers continue to study how these forces work together to control the Antarctic sea ice extent.

**Current Status and Trends**: In September 2008, Antarctic sea ice peaked at 18.5 million square kilometers), slightly below the monthly average for 1979-2000. The February 2009 minimum of Antarctic sea ice was also slightly below average, at 2.9 million square kilometers. Antarctic sea ice peaks in September, and reaches a minimum in February. In some places, sea ice melts completely in the summer. Since 1979, the total annual Antarctic sea ice extent has increased about 1 percent per decade. Compared to the Arctic, the signal has been a "noisy" one, with wide year–to-year fluctuations relative to the trend. The largest summer minimum in the satellite record occurred in February 2003. The largest winter maximum occurred in September 2006. The 2006 maximum was interesting because it followed a February minimum that was the third *lowest* on record.

Unlike the Arctic, where the downward trend is consistent in all sectors, in all months, and in all seasons, the Antarctic picture is more complex. Based on data from 1979-2006, the annual trend for four of the five individual sectors was a very small positive one, but only in the Ross Sea was the increase statistically significant (greater than the natural year-to-year variability). On the other hand, ice extent *decreased* in the Bellingshausen/Amundsen Sea sector during the same period. The variability in Antarctic sea ice patterns in different sectors and from year to year makes it difficult to predict how Antarctic sea ice extent could change as global warming from greenhouse gases continues to warm the Earth. Climate models predict that Antarctic sea ice will respond more slowly than Arctic sea ice to warming, but as temperatures continue to rise, a long-term decline is expected.

You might wonder why the negative trends in Arctic sea ice seem to be more important to climate scientists than the smaller increase in Antarctic ice. Part of the reason, of course, is simply that the size of the increase is much smaller and slightly less certain than the Arctic trend. Another reason, however, is that the complete summertime disappearance of Northern Hemisphere ice would be a dramatic departure from what has occurred throughout the satellite record and likely throughout recorded history. In the Antarctic, however, sea ice already melts almost completely each summer. Even if it completely disappeared in the summer, the impact on the Earth's climate system would likely be much smaller than a similar disappearance of Arctic ice.

You might also wonder how Antarctic sea ice could be increasing, even a little bit, while groups are claiming that the global warming from greenhouse gases is raising the planet's average surface temperature. It's a question scientists are asking, too. Some argued that the reason may be that other atmospheric changes are softening the influence of global warming if any on Antarctica. For example, the ozone hole that develops over Antarctica each spring actually intensifies a perpetual vortex of winds that circles the South Pole. The stronger this vortex becomes, the more isolated the Antarctic atmosphere becomes from the rest of the planet. In addition, ocean circulation in the Antarctic behaves differently than it does in the Arctic. Around Antarctica, warm water moves downward in the ocean's water column, making sea ice melt from warm water less likely. However, this is not new but it was there in the past and will be there in the future as it is a natural system.

A new NASA study says that an increase in Antarctic snow accumulation that began 10,000 years ago is currently adding enough ice to the continent to outweigh the increased losses from its thinning glaciers.

The research challenges the conclusions of other studies, including the IPCC 2013 report, which says that Antarctica is overall losing land ice. According to the new analysis of satellite data, the Antarctic ice sheet showed a net gain of 112 billion tons of ice a year from 1992 to 2001. That net gain slowed to 82 billion tons of ice per year between 2003 and 2008.

We're essentially in agreement with other studies that show an increase in ice discharge in the Antarctic Peninsula and the Thwaites and Pine Island region of West Antarctica", says Jay Zwally, a glaciologist with NASA Goddard Space Flight Centre in Greenbelt, Maryland, and lead author of the study, which was published on October 30 of 2015 in the Journal of Glaciology. "Our main disagreement is for east Antarctica and the interior West Antarctica -- there, we see an ice gain that exceeds the losses in the other areas". Zwally added that this team "measured small height changes over large areas, as well as the large changes observed over smaller areas".

Mass changes of the Antarctic ice sheet impact sea-level rise as climate changes, but recent rates have been uncertain. Ice, Cloud and land Elevation Satellite (ICESat) data (2003–08) show mass gains from snow accumulation exceeded discharge losses by 82± 25 Gt $a^{-1}$, reducing global sea-level rise by 0.23 mm $a^{-1}$. European Remote-sensing Satellite (ERS) data (1992–2001) give a similar gain of 112 ± 61 Gt $a^{-1}$. Gains of 136 Gt $a^{-1}$ in the East Antarctica (EA) and 72 Gt $a^{-1}$ in four drainage systems (WA2) in the West Antarctic (WA) exceed losses of 97 Gt $a^{-1}$ from three coastal drainage systems (WA1); and 29 Gt $a^{-1}$ from the Antarctic Peninsula (AP). EA dynamic thickening of 147 Gt $a^{-1}$ is a continuing response to increased accumulation (>50%) since the early Holocene. Recent accumulated losses of 11 Gt $a^{-1}$ in EA indicates thickening is not from contemporaneous snowfall increases. Similarly, the WA2 gain is mainly (60 Gt $a^{-1}$) dynamic thickening. In WA1 and the AP, increased losses of 66 ± 16 Gt $a^{-1}$ from increased dynamic thinning from accelerating glaciers are 50% offset by greater WA snowfall. The decadal increase in dynamic thinning in WA1 and the AP is approximately one-third of the long-term dynamic thickening in EA and WA2, which should buffer additional dynamic thinning for decades.

**Antarctic Sea Ice Loss**: A study on warming of West Antarctica since the 1957 geophysical year correlates widespread warming in West Antarctica and sea ice decline. Whether sea ice decline has led to warming temperatures on the continent, or whether both phenomena are caused by something else is not currently known.

One concern related to potential Antarctic sea ice loss is that sea ice may stabilize Antarctic ice shelves. Ice shelves are slabs of ice that partly rest on land and partly float. Ice shelves frequently calves icebergs, and this is a natural process, not necessarily a sign of climate change. But the rapid disintegration and retreat of an ice shelf (such as the collapse of the Larsen B shelf in 2001) is a warming signal. Although sea ice is too thin to physically buttress an ice shelf, intact sea ice may preserve cool conditions that stabilize an ice shelf because air currents passing over sea ice are cooler than air currents passing over open oceans. Sea ice may also suppress ocean waves that would otherwise flex the shelf and speed ice shelf breakup.

**Icebergs are *not* sea ice**: they're floating fragments of glaciers or ice shelves. Iceberg formation (and ice shelf retreat) may increase in regions of Antarctica where sea ice is declining because sea ice protects ice shelves by cooling air temperatures and dampening waves. The interaction between sea ice loss and ice shelf retreat merits careful study because many ice shelves are fed by glaciers. When an ice shelf disintegrates, the glacier feeding it often accelerates. Because glacier acceleration introduces a new ice mass into the ocean, it can raise ocean level. While sea ice melt does not directly lead to sea level rise, it could contribute to other processes that do, both in the Arctic and the Antarctic. Glacier acceleration has already been observed on the Antarctic Peninsula, although the accelerating glaciers in that region have so far had a negligible effect on ocean.

**Ice Melt**: Global warming theory predicts that rising levels of carbon dioxide will generally warm the air and cause an increasing loss of sea ice. The sea ice is located mostly outside the Antarctica Circle and inside the Arctic Circle and thus it is Antarctica sea ice that should be the most sensitive indicator of global warming. Arctic ice is affected by too many other compounding factors, like physical impacts and natural cycles like AMO, PMO, 50-70 year, etc. This suggests less ice melt on an average and thus negative contribution to sea-level rise – see Alaska. The three coldest Arctic summers have all occurred in the last 5 years (2009, 2010 & 2013) since 1958. From 1980 – 2000, both AMO & PDO show warm phases, and so the Arctic melt was likely caused by the Warmer surface sea water. But now that the PDO has flipped to its cool phase and the AMO is transitioning to its cool phase, we can expect the Arctic to begin its recovery over the next 2 decades.

David Middleton [http://www.clipartbest.com] presented a report on 31$^{st}$ December 2015 in which he presented the Greenland ice loss in three different dimensions and **Figure 2.11** is one of them showing the loss relative to the total. According to this report Greenland has lost 9,000 billion tons of ice in a century, which means 8165 Gt and otherwise it is 8165 cubic kilometers. The ice volume at the onset of the 20$^{th}$ century should be 2,608,165 cubic kilometers spread over 1.7 million square kilometers. Thus the thickness of ice is 1.5 km or 1,500 m. Thus % surviving is 99.7% and lost only 0.3% in the 20$^{th}$ century. This means 5m thickness was lost in 1,500 m thick ice sheet.

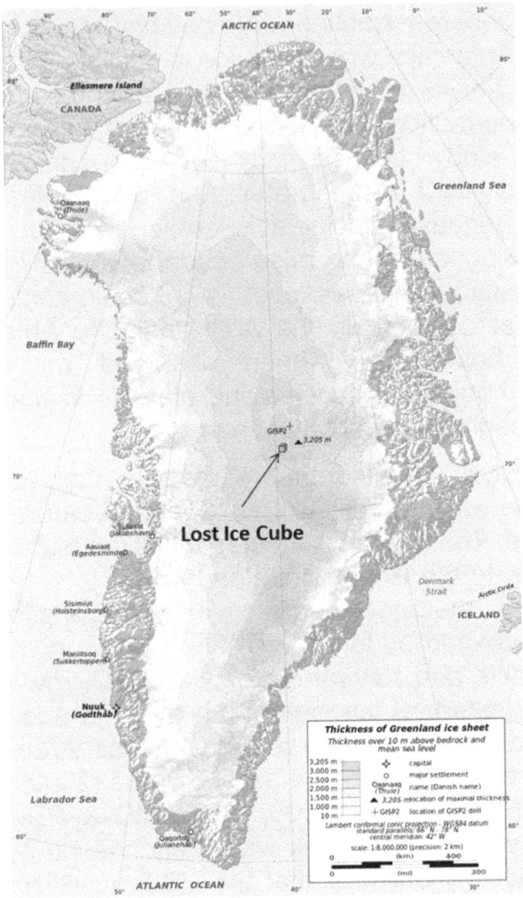


**Figure 2.11** Greenland ice loss in 20$^{th}$ century relative to total

### 2.3.2.2 Glaciers Melt

Indian scientists [including myself, wrote to ministry of forests & environment; and sent an article to CSE Down To Earth magazine, just before it is going for printing, announced Noble Prize to IPCC & Al Gore, and thus in its place another article of mine relating to Polavaram project was published] questioned the IPCC conclusion on Himalayan Glacier melt issue. R. K. Pachauri, the then Chairman of IPCC, dismissed criticism, claim it as "voodoo science". After 2009 December Copenhagen fiasco, IPCC says the Himalayan Glaciers won't melt by 2035 & expressed regret by saying that established standards of evidence not applied properly.

Also, Al Gore withdrew his conclusion: Greenland will become ice free in 5 years. IPCC & Al Gore received Noble Prize for such reports.

Some argued that "Glaciers are melting at unprecedented rates due to global warming, causing additional climate changes. About a quarter of the globe's glacial loss from 1851-2010, and approximately two thirds of glacial loss between 1991 and 2010, is attributable directly to global warming caused by human-produced greenhouse gases". According to the National Snow and Ice Data Center, global warming from human-produced greenhouse gases is a primary cause of the "unprecedented" retreat of glaciers around the world since the early 20th century. Since 1980 glaciers worldwide have lost nearly 40 feet (12 meters) in average thickness. According to a 2013 IPCC report, "glaciers have continued to shrink almost worldwide" over the prior two decades, and there is "high confidence" (about an 8 out of 10 chance) that Northern Hemisphere spring snow continues to decrease. If the glaciers forming the Greenland ice sheet were to melt entirely, global sea levels could increase by up to 20 feet. Melting glaciers also change the climate of the surrounding region. With the loss of summer glacial melt water, the temperatures in rivers and lakes increase. According to the US Geological Service, this disruption can include the "extinction of temperature sensitive aquatic species."

According to a 2013 study of ice cores published in *Nature Geosciences*, the current melting of glaciers in Western Antarctica is due to "atmospheric circulation changes" that have "caused rapid warming over the West Antarctic Ice Sheet" and cannot be directly attributed to human caused climate change. According to one of the study authors, "[i]f we could look back at this region of Antarctica in the 1940s and 1830s, we would find that the regional climate would look a lot like it does today, and I think we also would find the glaciers retreating much as they are today." According to Christian Schlüchter, Professor of Geology at the University of Bern, 4,000 year old tree remains have been found beneath retreating glaciers in the Swiss Alps, indicating that they were previously glacier-free. According to Schlüchter, the current retreat of glaciers in the Alps began in the mid-19th century, before large amounts of human caused $CO_2$ had entered the atmosphere.

In 2014 a study of 2,181 Himalayan glaciers from 2000-2011 showed that 86.6% of the glaciers were not receding [this was also informed to Indian Parliament by the minister of forests & Environment and climate change after his return from Paris meet in December 2015]. Heavy snowfall recording was reported in the last two years in Himalayan zones.

There are several local and regional causes for ice melt such as smoke from religious site, physical impacts cause melting or destruction of ice: sporting, pilgrimage, deforestation with increased population growth, infrastructure development, mountaineering, etc. In the Arctic and Antarctica zones in addition there are several other activities like drilling for gas and oil and Earthquake and Volcanic activities affect the ice melting or ice destruction.

Geological Survey of India monitoring few important glaciers in Himalayan region. Gangotri, is one of them, feed the main river Ganga. Due to formation of a fault zone the ice started receding and now it started recovering.

#### 2.3.2.3 Sea Level Change

World's ice cap and glaciers contain about 30,000,000 cubic kilometers of ice, of this 76.8% is in East Antarctica Ice Sheet. 10.0% in West Antarctica Ice Sheet, 9.8% in Greenland Ice caps of which 97.9% is grounded ice and 2.1% floating ice. Globally ice (both grounded and floating) represents about 2.0% of the world's water. Wm Robert Johnson says – Greenland and Antarctica would take thousands of years to melt. Antarctica ice cap appears currently growing rather than shrinking. Greenland's ice cap would take 2,000 to 10,000 years to disappear. However, at present it shows clear cut seasonal and annual variations within the natural cyclic variation in sea/ocean temperature. The theory: with the global warming Arctic and Antarctic ice is melting and thus causing the sea level rise. Hydrological cycle: the evaporated water from Oceans, water bodies comes back as rain and snow. Thus each year about 8 mm of water goes into the Antarctica and Greenland ice sheets as snowfall. If no ice returned to the Oceans, sea level would drop by 80 mm every 10 years. Scientists don't know which is greater, the ice going in or the ice coming out. The predictions of sea level rise are presented with model simulations based on global warming predictions and predicting the low lying islands, coastal cities will disappear and thus its impact on people of coastal zones.

According to the IPCC, there has been a "substantial" human contribution to the global mean sea-level rise since the 1970s, and there is "high confidence" (8 out of 10 chance) that the rate of sea-level rise over the last half century has accelerated faster than it has over the previous 2,000 years. A 2006 study found that "significant acceleration" of sea-level rise occurred from 1870 to 2004. Between 1961 and 2003 global sea levels rose 8 inches. An October 2014 study published in the *Proceedings of the National Academy of Sciences* concluded that the rate of sea level rise over the past century is unprecedented over the last 6,000 years. A separate October 2014 study said that the global sea level is likely to rise 31 inches by 2100, with a worst case scenario rise of 6 feet. Climate Central predicts that 147 to 216 million people live in areas that will be below sea level or regular flood areas by the end of the century if human-produced greenhouse gas emissions continue at their current rate.

Wikipedia writes that sea level rise has been "estimated" to be on average +2.6 mm and +2.9 mm per year ± 0.4 mm since 1993. Additionally, sea level rise has accelerated in recent years. Over the 21st century, the IPCC projects for a "high emissions scenario", that global mean sea level could rise by 52–98 cm. The IPCC's projections are conservative, and may underestimate future sea level rise. Other estimates suggest for the same period that global mean sea level could rise by 0.2 to 2.0 m (0.7–6.6 ft), relative to mean sea level in 1992. Widespread "coastal flooding would be expected" if several degrees of warming is sustained for millennia. For example, sustained 'global warming' of more than 2 °C (relative to pre-industrial levels) could lead to eventual sea level rise of around 1 to 4 m due to thermal expansion of sea water and the melting of glaciers and small ice caps. Melting of the Greenland ice sheets could contribute an additional 4 to 7.5 m over many thousands of years. It has been estimated that we are already committed to a sea-level rise of approximately 2.3 meters for each degree of temperature rise within the next 2,000 years. Warming beyond the 2 °C target would potentially lead to rates of sea-level rise dominated by ice loss from Antarctica. Continued $CO_2$ emissions from fossil sources could cause additional tens of meters of sea level rise, over the next millennia and eventually ullimately eliminate the entire

Antarctic ice sheet, causing about 58 meters of sea level rise. All these are hypothetical and far from reality. They help in sensationalization.

Sea levels have been steadily rising for thousands of years, and the increase has nothing to do with humans. A 2014 report by the Global Warming Policy Foundation found that a slow global sea level rise has been ongoing for the last 10,000 years. When the Earth began coming out of the Pleistocene Ice Age 18,000 years ago, sea levels were about 400 feet lower than they are today and have been steadily rising ever since. According to Professor of Earth and Atmospheric Sciences at the Georgia Institute of Technology, Judith Curry, PhD "it is clear that natural variability has dominated sea level rise during the 20th century, with changes in the Ocean heat content and changes in precipitation patterns." Freeman Dyson, Emeritus Professor of Mathematical Physics and Astrophysics at the Institute for Advanced Study at Princeton University, has stated that there is "no evidence" that rising sea levels are due to anthropogenic climate change.

IPCC and other groups who are in global warming trap, advocating that sea level rise is due to global warming. However, when we look at the data of sea level changes over different parts of the globe, it does not support the global warming as the cause of sea level changes. There are wide variations in sea level changes with ocean to ocean and within each ocean. They are highly related to local conditions. With all this, Wikipedia reports that for the last 100 years, sea level has been rising at an average rate of about 1.8 mm per year.

**Table 2.3** presents the Sea Level Changes over different parts of the globe. On a global scale they varied between -17.59 to + 9.65 mm/year. Also the error variations over different regions of the globe and as well within a region are not uniform. The global average was estimated at 1.7 to 1.8 mm/year. However, it has no meaning with so much variation. Can we attribute such variations to global warming?

San Francisco Airport is in the Ocean waters and yet so far no change. In Italy along the coast, traditional and historical wells showed no change in sea level. Silt deposition from Ganga & Brahmaputra Rivers causing the submergence of Mangroves in Sunderbans and hutments along the Brahmaputra in Bangladesh. Destruction of Coastline for commercial activities and destruction of Mangroves allowed sea water entering into nearby villages and thus people are attributing it to sea level rise – La Nina & El Nino play an important role. In several parts, sea sand is lifted for other uses and thus causing the sea water entering in to nearby areas.

**Table 2.3** Sea level changes over different regions of the Globe

| Sea Level Trends | | |
|---|---|---|
| **Region** | **No. of Stations** | **Changes in Sea Level, mm/year** |
| **Atlantic Ocean and Seas** | | |
| Northwest Atlantic | 54 | -9.48 to 6.02 |
| Gulf of Mexico | 19 | 1.73 to 9.65 |
| Caribbean Sea | 8 | 1.41 to 5.31 |
| Norwegian Sea | 9 | -2.06 to 2.33 |
| North Sea | 15 | -3.17 to 2.53 |
| Baltic Sea | 33 | -8.10 to 2.91 |
| Northeast Atlantic | 10 | 0.07 to 2.58 |
| Mediterranean Sea | 19 | 0.28 to 4.37 |
| Black Sea | 6 | 1.22 to 6.59 |
| Southwest Atlantic | 9 | 0.06 to 4.20 |
| **Pacific Ocean and Seas** | | |
| Northwest Pacific | 9 | -0.43 to 9.39 |
| Sea of Japan | 7 | -0.20 to 3.73 |
| East China Sea | 10 | 0.04 to 5.35 |
| South China Sea & nearby regions | 13 | -1.25 to 5.38 |
| Northeast Pacific | 57 | -17.59 to 4.32 |
| Southeast & Southwest Pacific | 8 | 0.09 to 2.45 |
| **Pacific Islands** | | |
| North | 16 | 0.60 to 7.23 |
| South | 12 | -2.59 to 6.30 |
| Indian Ocean & Seas | | |
| Indian Ocean | 19 | 0.32 to 4.67 |
| Southeast Atlantic & Indian Ocean Region | 4 | 0.60 to 2.39 |
| Arctic Circle & Southern Ocean | | |
| Arctic Circle Region | 9 | -2.25 to 3.17 |
| Southern Ocean Region | 2 | -4.82 to 1.43 |

*Source*: http://tideandcurrents.noaa.gov./sltrends/globalregional.html/

Willies Eschenbach presented an article on Lennox Island in Canada, a tiny Island off the Atlantic Coast. It is reported by others that this Island is shrinking as the climate warms and thus the seas rise. The fact says it is not so. Here we are not dealing with Islands made of rocks. Poor folks live on nothing more than a shifting pile of mud, silt and sand, located near the mouth of a river. The easternmost projecting point has been eroded away entirely, as has the southernmost. On the other hand, the northernmost part of the Island has grown by accretion of mud, silt and sand on the northern side, and has moved northwards while losing land on the south side. On an average maintain the size of the Island.

Open letter to President Maldives by Prof. Nils Axel Morner, Stockholm, Sweden, October 20, 2009 "You have recently held an undersea cabinet meeting to raise awareness of the idea that global sea level is rising and hence threatens to drown the Maldives. This proposition is not founded in observational facts and true scientific judgments, accordingly it is incorrect – had reminded quite stable for the last 30 years. In the last 2000 years, sea level has oscillated with 5 peaks reaching 0.6 to 1.2 m above the present sea level; from 1790 to 1970 sea level was about 20 cm higher than today; in the 1970s sea level fell by about 20 cm to its present level; sea level has reminds stable for the last 30 years, implying that there are no traces of any alarming on-going sea level rise. Maldives has a beautiful under water disco.

There are two types of islands/coast line, namely mud based and rock based. **Figures 2.12a, b, c & d & 2.13a & b** presents mud based island/coast erosion; and **Figure II.14** presents the sea level changes in Tuvalu islands, a rock based. In the former islands coast lines are eroded over a time but in the later the sea level showed a fluctuations [natural].

Sea level change follow the ice melt pattern only which in turn follow the 60-year cycle but Arctic and Antarctic ice melt follow opposite patterns. US West Coast shows a falling and the East Coast shows a rising. An article published in Nature Geosciences [July 2010] present that "Indian Ocean sea level change in warming climate" observed "while a number of areas in the Indian Ocean region are experiencing sea level rise, sea level is lowering in other areas" – as per the general circulation patterns. The study further indicates that the Seychelles Islands and Islands of Zanzibar's coast show the largest sea level drop.

Human direct impacts and natural disasters within the ocean floor and along the coast play vital role: volcanic activity, earthquakes, cyclonic activity, human developmental & other activities, and destruction of the coastal zones. Extraction of oil & gas and water creates land subsidence and thus creates sea level change like scenarios.

Sea Levels are affected by several factors and among these the natural variations play an important role. They are of two types, namely short lived and long lived ones. Human actions aggravated these changes through destruction of coastlines for settlements, digging for water, oil & gas; Tsunamis, earthquakes, cyclones, etc. [see an example **Figure 2.12a & b**]. Every time there is a storm, the North Sea continues its relentless erosion of the coast near the small English town of Happisburgh [**Figure 2.12a**]. The old coastal defenses are largely ineffective. Here, a Second World War bunker has fallen from the eroded cliffs, while elsewhere along the coast homes have already been lost to the sea. **Figure 2.12 b** presents breaches along the Visakhapatnam beach line.

**(a)**

VIZAG LURKING DANGER BEACH

N Kanaka

People stop to watch the beach in Vizag city which has witnessed rapid erosion over the last few weeks. On Wednesday, a portion of the beach along the bund near Beach Road got completely eroded, exposing the bund's wall which left residents living in nearby houses jittery

**(b)**

**(c)**

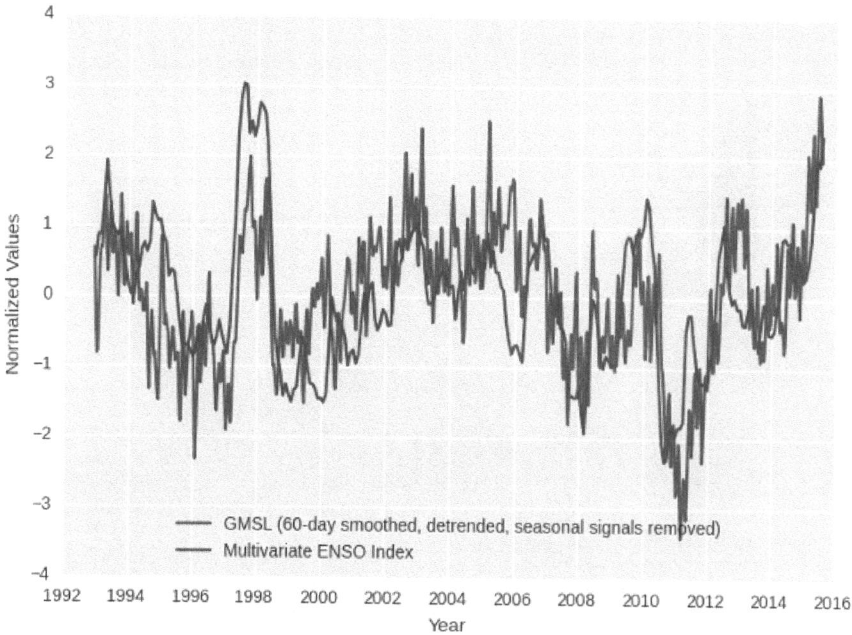


(d)

**Figure 2.12** (a) Erosion of the coast near the small English town of Happisburgh, (b) Visakhapatnam Beach Erosion in India, (c) Pacifica/San Francisco & (d) Sea level versus El Nino events

(a)

[b]

(a) A boy from Burhanuddin sub-division of Bhola island district by the banks of the Meghna river. (b) People in the area say erosion and inundation has aggravated in the recent past

**Figure 2.13** Island Erosion in Bangladesh [a & b]

Jim Steele [31 January2016] observed that "Ill informed in 1949, developers constructed several homes at the top of a 20-meter sea cliff along Esplanade Drive in the City of Pacifica. During heavy winter storms of the 1997/98 El Nino, 10 m of local coastline were rapidly eroded, eliminating the coast vestiges of the backyards that had survived the 1982 El Nino – see Pre-1997 photograph, **Figure 2.12c**. Andres Valencia responded Dr. Steele you wrote "sea level rise varies most between El Nino and La Nina events", to illustrate this see **Figure 2.12d.**

Ice is confined to outside the South Polar Ring and inside the North Polar Ring. That means South Polar ice melt is the true reflection of global warming impact on ice melt. At present it is not showing any melt in the Southern Polar zones. The Southern Polar zones are on the contrary building ice. North Polar zones are losing the ice but this is within the long-term standard deviation around the mean – within the accepted statistical terms. In the North Polar zone, impacts other than climate are also contributing to ice melt.

Alaska shows a large fall in sea level. Along the USA coastlines, a large part showed a 0 to 1.0 feet fall, this may be associated with human activity along the coastal zones, tidal erosion, etc. Also, error variations are far higher than the estimated rise, which is not statistically significant. In addition, all these localized natural variations play a vital role – even the global temperature showed a 60-year cycle – sine curve. Cyclonic activities including Hurricanes and Typhoons – and precipitation, all present cyclic variations. With the growing population of the planet and building more structures in the path of cyclonic storms – that includes Hurricanes & Typhoons – and Tornadoes, and thus this makes ordinary storms more damaging.

Reports state that sea levels are rising at an unprecedented rate due to global warming. As human-produced greenhouse gases warm the planet, sea levels are rising due to thermal expansion of warming ocean waters as well as melt water from receding glaciers and the polar ice cap. Experts differed in their opinions about the extent to which specific factors play a role here, namely [1] between 15 and 50% attributed to the temperature related expansion of sea water; [2] between 25 to 45% to be caused by melting of mountain glaciers outside polar region; and [3] between 15 and 40% is ascribed to the melting of the Greenland and Antarctic ice cap. We can see the wide ranges in all the three types, meaning they are only speculative.

All these are possible only when global warming is as they predicted by modelers. If we look at **Figures 2.15 & 2.16,** it shows natural variations in polar ice and ocean temperature patterns. In the case of polar ice, Arctic Sea Ice Extent shows a reverse pattern at Antarctic Sea Ice Extent [**Figure 2.7**]. Also this is seen in standardized temperature [**Figure 2.17**]. **Figure 2.17** presents a clear cut 60 year cycle in Pacific.

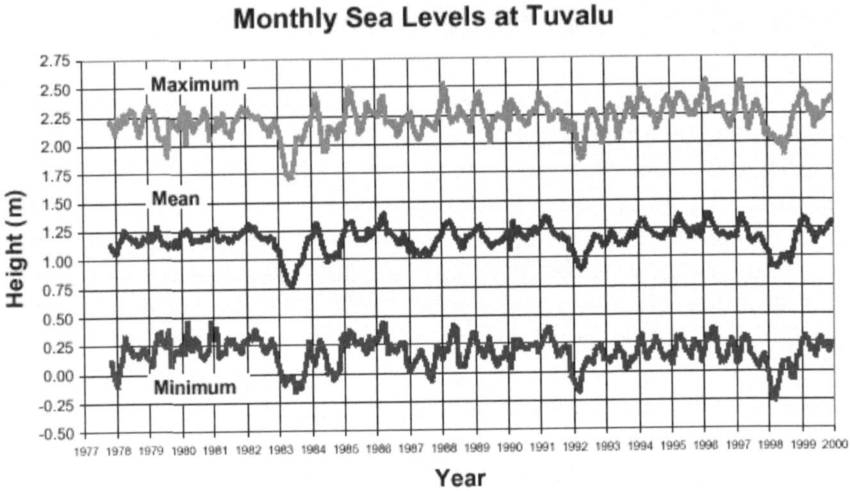


**Figure 2.14** Monthly sea levels at Tuvalu

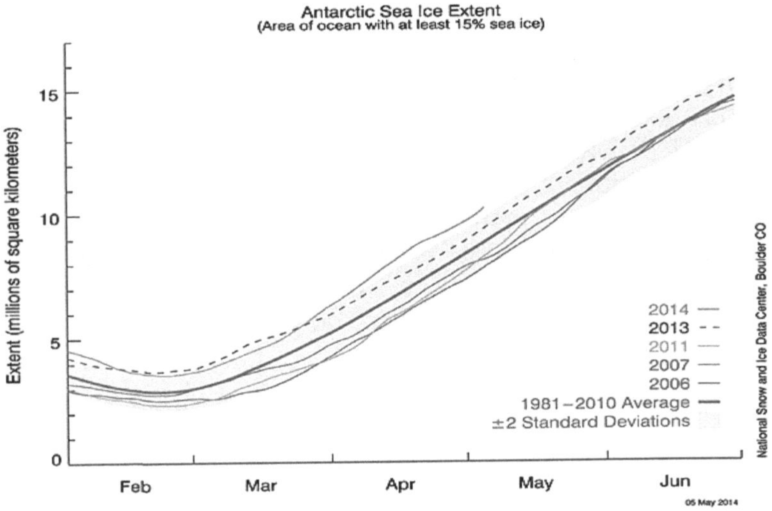


(a)

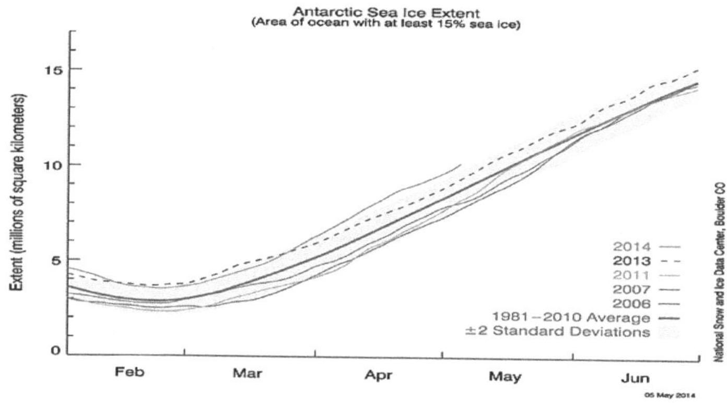


Sea ice extent in recent years (in million km$^2$) for the northern hemisphere, as a function of date.

(b)

**Figure 2.15** [a] Arctic and [b] Antarctic sea ice extent

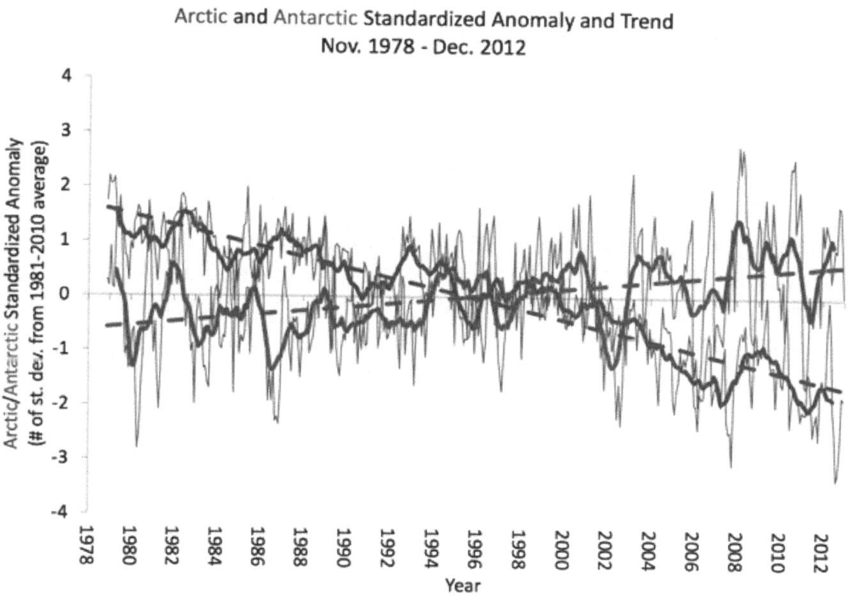


**Figure 2.16** Arctic and Antarctic standardized temperature anomaly and trend

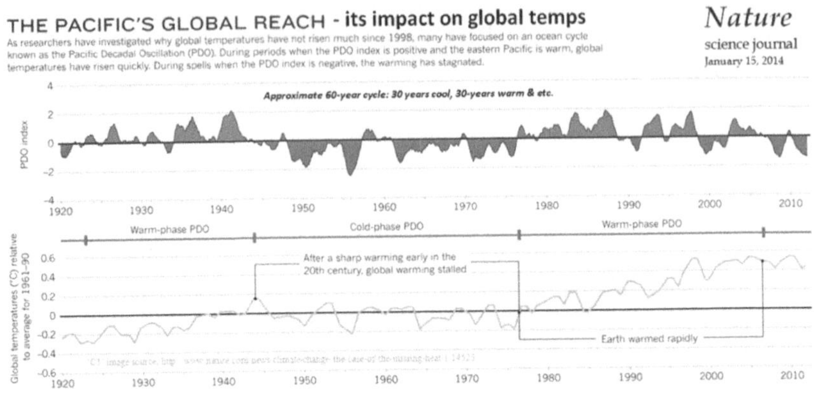


**Figure 2.17** The Pacific's Global Reach – its impact on global temperature

## 2.3.3 Ocean Acidity

Reports suggest that "Ocean acidity levels are increasing at an unprecedented rate that can only be explained by human activity. As excess human-produced $CO_2$ in the atmosphere is absorbed by the oceans, the acidity level of the water increases. Acidity levels in the oceans are 25-30% higher than prior to human fossil fuel use". According to a 2014 US Government Accountability Office (GAO) report, oceans have absorbed about 30% of the $CO_2$ emitted by humans over the past 200 years, and ocean acidity could rises approximately 100-200% above preindustrial levels by 2100. According to a 2013 report from the World Meteorological Organization, the current acceleration in the rate of ocean acidification "appears unprecedented" over the last 300 million years. High ocean acidity level threatens marine species, and slows the growth of coral reefs. According to a 2014 report by the Convention on Biological Diversity, "it is now nearly inevitable" that within 50-100 years continued human produced $CO_2$ emissions will increase ocean acidity to levels that "will have widespread impacts, mostly deleterious, on marine organisms and ecosystems."

The acidity levels of the oceans are within past natural levels, and the current rise in acidity is a natural fluctuation, not the result of increased levels of Carbon Dioxide in the Atmosphere. The pH of average ocean surface water is 8.1 and has only decreased 0.1 since the beginning of the industrial revolution (neutral is pH 7, acid is below pH 7). In 2010 *Science* published a study of ocean acidity levels over the past 15 million years, finding that the "samples record surface seawater pH values that are within the range observed in the oceans today." Increased atmospheric $CO_2$ absorbed by the oceans results in higher rates of photosynthesis and faster growth of ocean plants and phytoplankton, which increases pH levels keeping the water alkaline, not acidic. According to a 2010 paper by the Science and Public Policy Institute, "our harmless emissions of trifling quantities of carbon dioxide cannot possibly acidify the oceans."

### 2.3.4 Ocean Temperatures

Reports state that "Ocean temperatures are rising at an unprecedented rate due to global warming, and are causing additional climate changes". The IPCC stated in a 2013 report that due to human-caused global warming, it is "virtually certain" (99-100% probability) that the upper ocean warmed between 1971 and 2010. An October 2014 *Nature Climate Change* study said that the oceans are the "dominant reservoir of heat uptake in the climate system." A separate October 2014 study found that the oceans absorb more than 90% of the heat generated by human-caused global warming. Since 1970 the upper ocean (above 700 meters) has been warming 24-55% faster than previous studies had predicted. A May 2013 study published in *Geophysical Research Letters* found that between 1958-2009 the rates of warming in the lower ocean (below 700 m) "appear to be unprecedented'. According to an October 2013 study, the middle depths of the Pacific Ocean have warmed "15 times faster in the last 60 years than they did during apparent natural warming cycles in the previous 10,000". Warmer ocean waters can harm coral reefs and impact many species including krill, which are vital to the marine food chain and which reproduce significantly less in warmer water. Warming oceans also contribute to sea level rise due to thermal expansion, and warmer ocean waters can add to the intensity of storm systems. If this is true, this counters the acidification theory discussed above as with higher temperatures oceans release more $CO_2$ in to the atmosphere.

Deep ocean currents, not human activity, are a primary driver of natural climate warming and cooling cycles. Changes in ocean currents are primarily responsible for the melting Greenland ice sheet, Arctic sea ice, and Arctic permafrost. Over the 20th century there have been two Arctic warming periods with a cooling period (1940-1970) in between. According to a 2009 study in *Geophysical Research Letters*, natural shifts in the ocean currents are the major cause of these climate changes, not human-generated greenhouse gases. According to William Gray, PhD, Emeritus Professor of Atmospheric Science at Colorado State University, most of the climate changes over the last century are natural and "due to multi-decadal and multi-century changes in deep global ocean currents." Global cooling from 1940 to the 1970s, and warming from the 1970s to 2008, coincided with fluctuations in ocean currents and cloud cover driven by the Pacific Decadal Oscillation (PDO) - a naturally occurring rearrangement in atmospheric and oceanic circulation patterns. According to a 2014 article by Don Easterbrook, PhD, Professor Emeritus of Geology at Western Washington University, the "PDO cool mode has replaced the warm mode in the Pacific Ocean, virtually assuring us of about 30 years of global cooling, perhaps much deeper than the global cooling from about 1945 to 1977."

### 2.3.5 Wild Fires

IPCC attributes wildfires to extreme temperatures associated with global warming. Australian Prime Minister says there is no relationship between wildfires and global warming. In Australia the rainfall is highly variable and in association with this the temperature presents high year to year variations. Wildfires occur in dry seasons. Weather/climate controls fire activity and the success and failure in suppression efforts, very good at putting out fires in wet years, but very poor at putting out fires in dry ones – low humidity and high winds help fire spreading more areas.

### 2.3.6 Heat Related Mortality

From the AR-5 extremes are not part of the global warming so far & natural variations play vital role in coming two decades. J.R. Christy [submissions to house committee on science, space and technology on 2$^{nd}$ February 2016, US] presented that “In terms of heat waves [**Figure 2.19**], the number of days with 100 $^{o}$F observed in US from a controlled set of weather stations, hot days haven’t increased but in most recent years there has been a relative dearth of them”. So also satellite data presents for lower temperatures over the predicted ones at global level. In this testimony he also observed on the claim that satellites do not measure temperature: "In reality, the sensors on satellites measure by emitted radiation the same method that a physician uses to measure body temperature to high precision using an ear probe. Atmosphere oxygen emits microwaves, the intensity of which indirectly proportional to the temperature of the oxygen, and thus atmosphere. This data matches with balloon data."

Mortality due to cold exceeds the mortality due to heat 4 to 1. Britain experiences much higher death rates than countries in Scandinavia which have much colder winters but where there has been huge investment in insulating homes. Spiraling energy bills and poor insulation contributed to 24,000 deaths last winter which was relatively mild winter, as many elderly people cut back on their heating. The number will be even higher this year because of further increases in energy bills. Majority of the victims were over 75, exacerbate a number of underlying medical conditions in the elderly, mostly strokes and heart attacks.

Adulterated and polluted food is the major cause for human ill-health. The diseases like Malaria are confined to some specified zones & conditions over the globe. Climate change is not the cause for human ill-health but related to Pollution – air & water and chemical input agriculture technology – food pollution.

## 2.4 Concluding Remarks

From the discussions it is clear that we have neither quality global average temperature data nor greenhouse gases data nor a correct procedure to link warming to carbon dioxide increase in the atmosphere after industrialization. IPCC was using trial and error approach for fixing the sensitivity value. Even with this, it presents wide range from the mean [upper and lower limits present ±50% over the mean], which is invalid science. And yet warmist groups are postulating theories, day in and day out on ice melt, sea level rise, etc, that are creating false alarm. Human induced amplification of the risk of a hazard is also attributed to unproven global warming. Temperature presented high location-region specificity, land-ocean specificity, hemisphere specificity, climate system-general circulation system specificity. The changes in nature also presented such specificity with that civilizations have disappeared over time.

Weather presents wide variations from event to event over space and time and climate is the average of such variations. When we look at weather as a part of general circulation and oscillations within those zones, we get meaningful inferences. If we divert the discussion with "zero" global warming the forecasting system will die its natural death, this is exactly what is happening now in weather forecasting. Meteorologists world over must keep away from warmists tendencies. We know the fact on ozone depletion with the formation of circumpolar vortex but while interpreting the data we forget this fact. If we look in to past weather these are clearly evident all over the world. We rarely look in to past. Unfortunately warmists consider meteorology a hoax and only global warming is science.

With the growth of population large modifications took place to climate system, which influence the local and or regional weather systems. It has become a fashion to attribute changes associated with such system to global warming, a term which till to date not quantified. There are, however, two different human induced components that are in real time affecting the nature in wide variety of ways at local and or regional scales. Local governments must put emphasis on such systems that affect the respective environments.

Let me give an example [**Reddy, 2000**] on this: Orography plays major role on Indian rainfall. For example, Western Ghats help in producing wet areas on wind ward direction [Western parts of Ghats] and dry areas on leeward direction [Eastern parts of Ghats] during the Southwest Monsoon season. The opposite pattern is producing during the Northeast Monsoon Season – which gets copious rains with cyclonic activity in Bay of Bengal. The box effect of Himalayan Mountain ranges of the northeastern zone provides copious rains – the highest rainfall in the World is received at Chirapunji of this zone. Thus, the destruction of orography plays the major role in changing the rainfall around that zone. In Mumbai/India met stations are located in Colaba and Santacruz [airport]. Historical average rainfall at Santacruz is higher than Colaba by about 300 mm. For the purpose of expanding Santacruz airport for a separate international airport [I was a trainee on weather forecasting, staying inside the airport for a month] cut a hillock in the windward direction of the Southwest Monsoon on the Eastern and Northeastern side of the Santacruz met observatory on the runway. This results a decreasing trend in

Santacruz rainfall relative Colaba rainfall trend. With the international terminal in operation, high rise buildings have come up that acted as hillock and then reversed the decreasing trend in Santacruz rainfall now.

Climate scientists [?] have constructed models to predict what the Earth's climate will look decades, even hundreds of years in the future. Unfortunately, many major components of the Earth's climate system have not been accurately ascertained for very long. This makes such predictions suspect if not laughable. A case in point is variations in ocean circulation and temperature. In the Atlantic there is cycle for sea surface temperatures variation called the Atlantic multi-decadal oscillation [AMO]. The AMO is linked with decadal scale climate fluctuations like European summer precipitation, rainfall in Europe and India, Atlantic Hurricanes and variations in global temperatures. A new study in the Journal Nature reports that the AMO is again transitioning to a negative phase, meaning the vaunted "pause" in global warming may be with us for decades. In fact, scientists at the University of Southampton predict that cooling in the Atlantic Ocean could cool global temperature a half a degree Celsius.

Climate scientists and Oceanographers have studied ocean circulation patterns for years and – given the ocean's massive capacity for absorbing, storing, and releasing heat energy – they have long suspected linkage between using sea level along the east coast of the US to estimate sea surface temperature [SST] and climate variation. Now that linkage is the subject of a new study, titled "Ocean impact on decadal Atlantic climate variability as revealed by sea-level observations". In it, researchers from University of Southampton, led by Gerard D. McCarthy, have tried a new approach using sea level along the east coast of the US to estimate ocean circulation on decadal timescale. Here is the overview from the article abstract:

There are a number of cycles in the Earth's short-term, decadal scale, climate variability – the Pacific Decadal Oscillation [PDO], the Atlantic Multi-decadal Oscillation [AMO], and so forth. When it comes to the AMO the authors claim that circulation records do not go back for enough to get a really good handle on its behavior. This is despite the fact that NOAA publishes time series data going back to 1856. The researchers state their case this way.

The difficulty in linking ocean circulation changes to decadal climate variations lies in the fact that long observational records of ocean transports are rare. Measurements such as those of the Florida Current since 1982 and the Greenland-Scotland ridge transports since the mid-1990s are some of the longest continuous ocean transport records available. Continuous, full-depth, basin-wide measurements of the Atlantic overturning circulation only began in 2004 with the RAPID monitoring project at 26 °N. None of these records are long enough to directly link ocean circulation with decadal climate variations such as the AMO.

The authors' claim that sea-level measurements from tide gauges provide an integrated measure of water column properties, offering time series of sufficient duration to study decadal ocean circulation variation. While using tide gauges to investigate ocean circulation is not new – an attempt to estimate Gulf Stream flow using tide gauges was made by R. B. Montgomery in 1938 – this is the first time the technique has been applied to the AMO.

It is amazing that climate scientists seem to have forgotten where the energy that drives the Earth's climate comes from the first place. The fact that McCarthy et al. have detected a trend in the AMO that indicates a future cooling trend only reinforces what solar scientists have been saying for several years. This is because the shift in the AMO is a result of the climate system changing, not a fundamental cause of climate change. It is a cyclic oscillation in a complex system driven by solar energy.

We have seen the confusion relating polar ice scenario. Here we must note few important components that present wide differences with reference to Antarctic and Arctic regions. Southern Hemisphere when compared to Northern Hemisphere is covered by less land area and more ocean areas. Land to ocean heating patterns is different as well the ocean temperatures are lower than land temperatures. Because of this during the 6 months with the Sun's energy, rise in temperature in general over Antarctica zone is lower. Also, under general circulation scenario, the South Polar Regions are cooled further by formation of Circumpolar Vortex more frequently. This natural system is behind the more area being occupied by ice in South Polar Zones. 76.8% of the World's ice cap and glaciers are in East Antarctica Ice Sheet, 10.0% in West Antarctica Ice Sheet and 9.8% in Greenland Ice Cap; and of these 97.9% is grounded ice and 2.1% is floating ice. The rest is in glaciers.

There is a need to characterize the factors associated with rise and fall of ice cover in each of the regions and this will facilitate to forecasting of the future scenario. If we simply harping on global warming we never reach the reality; and thus their implications on sea level changes over different parts of the globe.

There exists several factors that influence Ice changes and sea level changes and they vary from location to location. The major issues in this direction are the earthquakes, volcanoes, human actions & extra-terrestrial actions relating to physical impacts, etc. The first and foremost thing we must do is to look in to historical data before we pronounce a weather event to global warming.

**Most Important scenarios**

1. The warming theory in fact counters the acidification theory as it is clear from the historical & current temperature versus carbon dioxide pattern [**Figure 1.12**], wherein warmer water releases carbon dioxide in to the atmosphere and thus warming and acidification will not go together. The other important scenario is that the sea level rise, wherein sea level rise is caused by melting of polar ice masses. This way we are adding more water to the Oceans. Thus, this not only cool the Ocean waters but also dilutes the carbon dioxide concentration and thus counters the theory of acidification, though cooling allows more carbon dioxide absorption by oceans. The third scenario is the hiatus, wherein three theories were put forth and one such theory is that warming is hiding in the oceans. If the warming is hiding in the oceans, it releases more carbon dioxide and thus more global warming. The fourth scenario is the natural multi-decadal and Southern Oscillation impacts. The fifth scenario is filth over ocean. Unless all these are integrated, the acidification & warming theories are of little significance.
2. Figure 2.18 presents an actual way of averaging for a calendar year from the Southern Hemisphere and Northern Hemisphere monthly averages. The basic question we should ask the IPCC and other UN agencies, how this average influences the nature? Impacts on nature are season, region & hemisphere specific. When we average

January month of NH with June month of SH and goes on to average of December month of NH with July month of SH, how they cause heat or cold waves; drought or floods; ice melt or sea level change; unusual events? All these are built on false foundations.

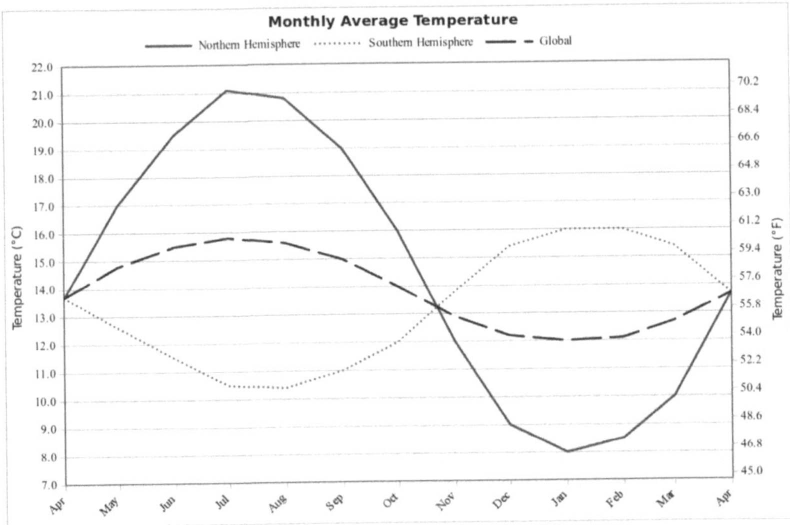


**Figure 2.18** Monthly average temperatures for both the hemispheres and the globe

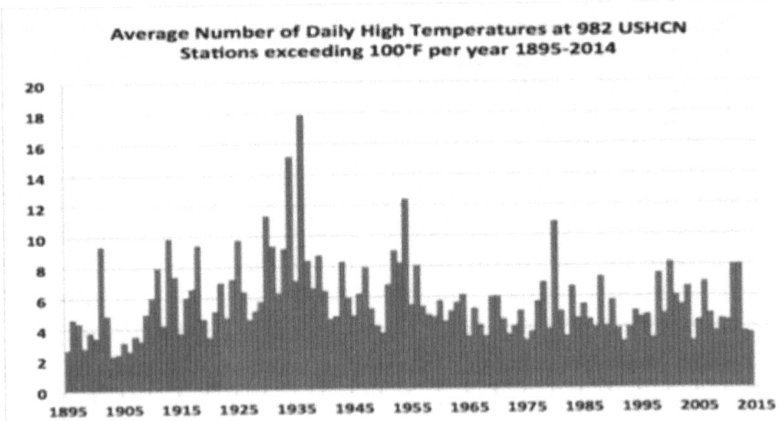


Above: Average number of days per-station in each year reaching or exceeding 100°F in 982 stations of the USHCN database (NOAA/NCEI, prepared by JRChristy).

**Figure 2.19** Average number of daily high temperatures during 1895-2014 in US

3. Abbott [Australian Prime Minister] considered investigation into 'exaggerated' Bureau of Meteorology temperature data. It drives me mental that these sorts of statements go unaddressed, says Sofhie Lewis from the Australian Research Council's Centre for Excellence in Climate Science. Lewis put in to an article "Tony Abbott's climate claims debunked: research dissects 2013 statement" [24th January 2016]. The report states that "To test if that might be the case Lewis ran a series of climate models in which the greenhouse effect was removed – so all that was left was natural variability. Unsurprisingly, in those models, high temperature records were less common than they are in reality. In other words, the record-breaking that we have seen cannot be explained by natural variation." On this I responded as: [1] IPCC noted more than the half of global average temperature anomaly is associated with the greenhouse effect component [this includes human component namely global warming and non-human component related to volcanoes and other aerosols plus filth component] and the rest to non-greenhouse component [mainly human induced component]. That means even after removing the global warming component, still there are two other components along with the systematic variations components [fluctuations and as well irregular variations]. The so called models in fact haven't taken these in to account. The difference between 2014 and 2015 is 0.18 °C [but they are different under different data sets and under satellite data sets]. The predicted rise associated with the El Nino is about 3 °C. With every peak of El Nino, on either side two depressions will occur and with this 2016/17 will automatically follow the hiatus uninterruptedly.
4. Eric Worral [24 July 2016] observed that "Remember all the recent press about the mild winter, how the bears were waking up early, flowers blooming in December, all a sign of the coming global warming apocalypse? Now that winter has turned cold and snowy --- When you have a theory which covers anything from 1.5 °C to 4.5 °C [or more!] temperature rise per doubling of $CO_2$, and when you claim with a straight face, that the utter failure of your theory, on terms which you yourself defined, doesn't invalidate it, when you have massively adjust the data to get the result you want, your

faith in the climate religion is not going to be troubled by the odd fury of snow".

5. If we look at the impacts on nature at local-regional scale activities: Himalayan Glaciers showed 86.6% of 2181 glaciers, not receding; Greenland ice [under clear cut seasonal variation], summer minimum reached the lower value in 2012 and now moved up – this is part of natural variation cycle; Seal level presented a variation -17.59 to 9.65 mm/yr over different parts, presenting a location-region specificity; droughts-floods and heat-cold waves, etc show a highly location-region specificity.

3 Chapter

# Natural Variability: Ground Realities

## 3.1 Introduction

Climate variations happen on all time scales, from one year to the next, as well as from one decade, century or millennium to the next; as well as on all spatial scales, from the local to regional to the global. The complex nature of this variability is a major obstacle to the reliable identification of global changes brought about – in the past, present or future – by the presence and activities of humanity on this planet. As an example, the oceanic El Nino phenomenon is most pronounced in the tropical Pacific off the coast of Peru, but the associated Southern Oscillation (ENSO) in the atmosphere has far-reaching, nearly global implications. It has a quasi-biennial component, with a characteristic recurrence time of 2–2.5 years, and a low-frequency one, with a recurrence of 4–5 years. Additional forms of inter-annual and inter-decadal variability in higher latitudes are associated with the North Atlantic Oscillation, the Pacific Decadal Oscillation, and the Arctic Oscillation; etc.

If you were to look at periods of natural variability over the past one million years, you would be likely focusing on the natural variability caused by the Milankovitch cycle. Those are the cycles that alter the shape of our orbit around the Sun as well as the tilt and wobble of the Earth axis. These cycles have periodicity in the scale of thousands of years with the average longer term period of around 100,000 years. These are explained with data derived indirectly using sea sediments, etc. The objective here is not to go in to indirectly derived data based variability. However, such cycles on short term show a linear trend [either increasing or decreasing based on the upward or downward part of the cyclic pattern]. We rarely account this.

Regional activities begin to increase in importance the shorter the time scale. These are explained directly with the observed meteorological data and with the tree rings data constructed indirectly. Short term natural variability (months, weeks, days, hours, & minutes) relates to the weather. Weather, as we all know, is the most variable of all these considerations. In short, variability happens. There will always be variability. Do humans influence variability? The answer is "Yes". The real difference is that the human influence changes the path of variability. In other words, variability will still happen, but on a different path. Natural variability will not go away, but certain forcing components will drive us in another direction. Among these the important one for understanding and for planning water resources availability and agriculture are short term variability. **WMO [1966]** presented a manual to separate man induced trend from natural variability. However, to understand the natural variability, it is essential to know about atmospheric general circulation patterns over different parts of the globe. The intricacies in natural variability will be provided by the atmospheric general circulation patterns. Also, extra-terrestrial phenomena do play on the natural variability in meteorological parameters. This includes both the Sun and the Moon.

## 3.2 Atmospheric General Circulation

### 3.2.1 Circulation Patterns

Climate is governed by the general circulation of the atmosphere, known as General Circulation System [**GCS**]. They are highly regional specific. In this climate system also plays vital role in modifying the GCS in a variety of ways. The world-scale systems of pressure and winds persist throughout the year or recur seasonally. Such winds transport heat from tropical to polar latitudes, thus maintaining the present patterns of world temperatures. The global circulation is driven by intense differences in insulation between tropical and polar-regions, and is strongly influenced by the Coriolis force associated with the rotation of the Earth.

Air moves vertically along the meridians and horizontally with the wind systems, both at ground level and in the upper atmosphere. Features of the general circulation of the atmosphere system represents the time-averaged pressure systems over the entire earth's surface as well the upper air. These include, Inter-tropical Convergence Zone [**ITCZ**], where in northeast and southwest trade winds at the equatorial lows converge and often results in calm winds, termed as "doldrums"; and others such as: Northeast trade winds, Southeast trade winds, Westerlies, Polar front, Hadley cell, Subtropical highs (historical maritime terminology called these areas the "horse latitudes"), Sub-polar lows--these are the mid-latitude extra-tropical cyclones, etc. All these are regionalized aspects.

Similarly the GCS is also influenced by longitudinal factor associated with land and water. Thus, **hot air rises in the tropics, moves north or south descends and returns in the equatorial "Hadley cells". Its path along the surface is bent into the trade winds by the Earth's rotation (Coriolis Effect). Two other cells in each hemisphere work similarly.**

Atmospheric circulation is the large-scale movement of air, and the means by which thermal energy is distributed on the surface of the Earth. This includes smaller ocean circulations also. **Figures 3.1a** presents the global general circulation pattern and **Figures 3.1b** presents the Hadley cell. The climate is the average of these systems and patterns – where and when they tend to occur again and again – it is stable over longer periods of time. **Figures 3.2a, b, c & d** presents the sea surface temperature departures for the [a] mean with El Nino episodes, [b] differences in measurements, [c] monthly sea surface temperature Nino 3.4 Index value and [d] oceanic Nino Index [OMI]. That means there existing large differences in the temperature measurements made by different groups as well there are large variations in El Nino and Southern Oscillation with year to year. Such scenarios make things complicated on GCS.

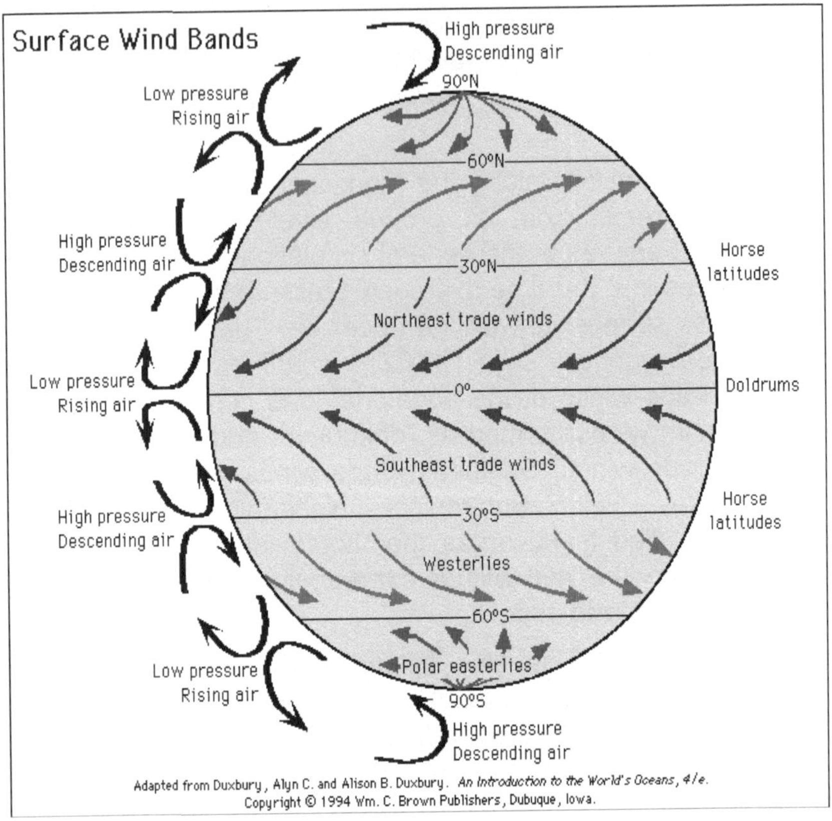


[a]

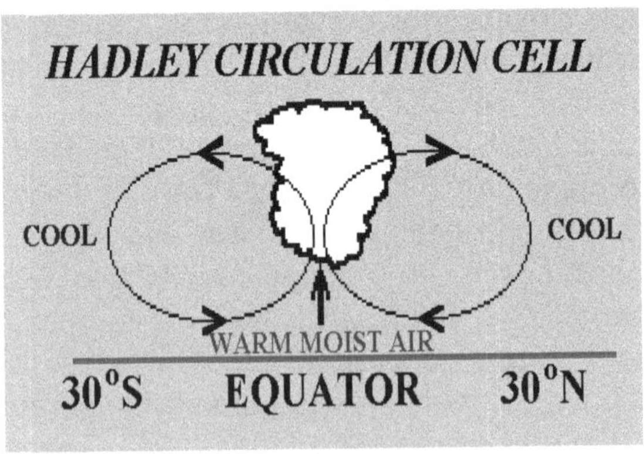


[b]

**Figure 3.1 [a]** Global general circulation pattern, & **[b]** Hardly circulation cell

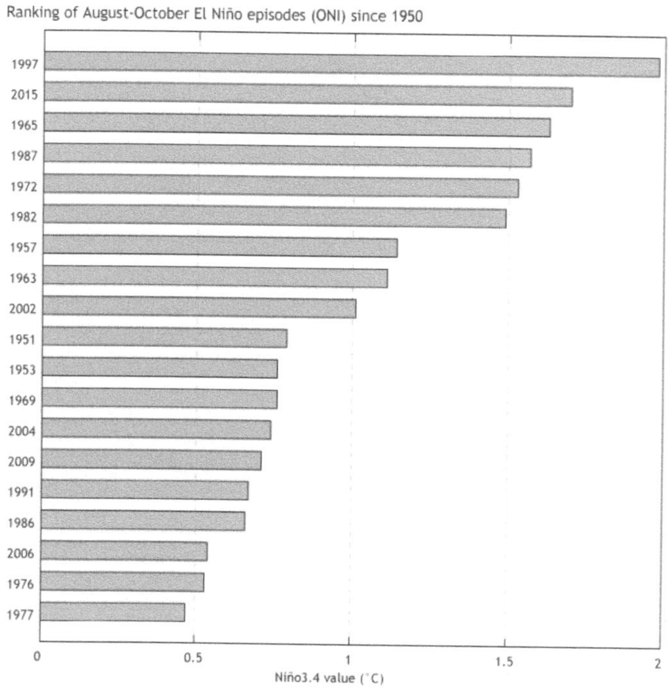


[a]

Ranking of August-October average sea surface temperature departures from the mean for all El Niño episodes since 1950. This measurement, the Oceanic Niño Index, uses ERSSTv4 data. Figure by climate.gov, data from CPC.

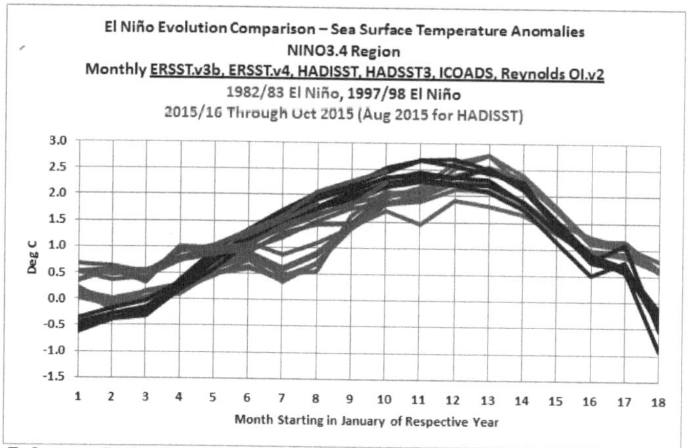


**Reference Period for Anomalies = 1981-2010** **Bob Tisdale**

[b]

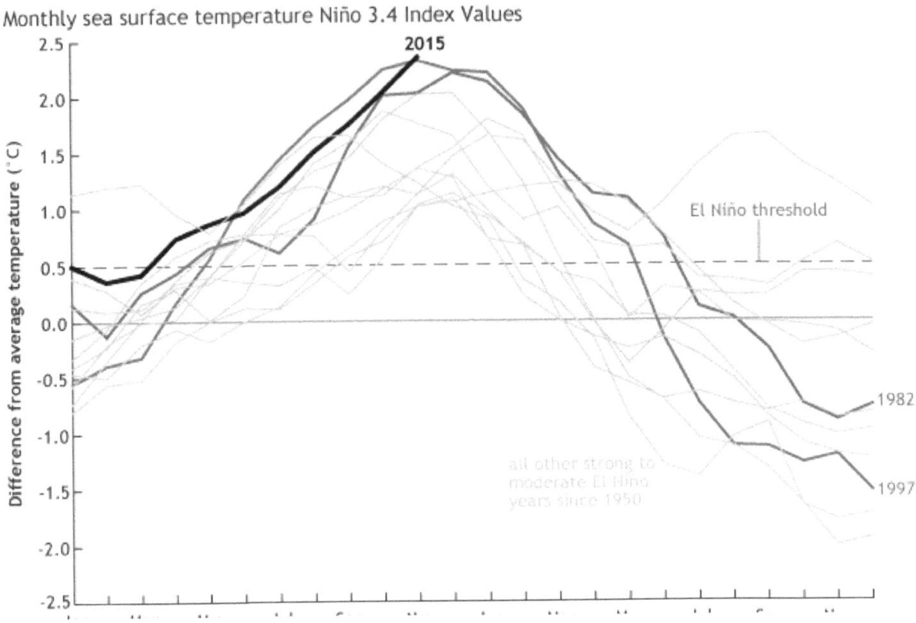


[c]

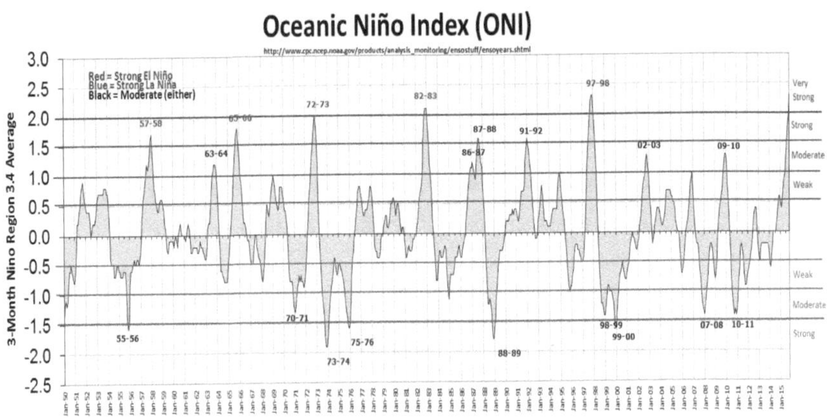


El Nino seasonal patterns

[d]

**Figure 3.2** [a] Ranking of August-October El Nino episodes [ONI] since 1950, [b] sea surface temperatures as measured by different groups for three El Ninos, [c] Monthly sea surface temperature Nino 3.4 Index values & [d] Oceanic Nino Index [ONI]

### 3.2.2 Latitudinal Circulation Features

The wind belts girdling the planet are organized into three cells, namely the Hadley cell, the Ferrel cell and the Polar ell [**Figure 3.1a**]. The vast bulk of the vertical motion occurs in the Hadley Cell [**Figure 3.1b**]. Low and high pressures on the Earth's surface are balanced by opposite positive pressure in the upper atmosphere.

The Hadley Cell mechanism is a closed circulation loop, which begins at the equator with warm, moist air lifted aloft in equatorial low pressure area, ITCZ, carried pole ward. At about 30° N/S latitude, it descends in a high pressure area. Some of the descending air travels equatorially along the surface, closing the loop of the Hadley Cell and creating the trade winds. Though the Hadley cell is described as lying on the equator, it is more accurate to describe it as following the sun's Zenith point, or what is termed the "thermal equator", which undergoes a semi-annual north-south migration. The Hadley system provides an example of a thermally direct circulation.

The Polar cell is likewise a simple system. Though cool and dry relative to equatorial air, air masses at the 60$^{th}$ parallel are still sufficiently warm and moist to undergo convection and drive a thermal loop. Air circulates within the Troposphere limited vertically by the tropopause at about 8 km. Warm air rises at lower latitudes and moves pole ward through the upper troposphere at both the north and the south poles. When the air reaches the polar areas, it has cooled considerably, and descends as a cold, dry high-pressure area, moving away from the pole along the surface but veering westward as a result of the Coriolis effect to produce the Polar easterlies.

The outflow from the cell creates harmonic waves in the atmosphere known as Rossby waves. These ultra-long waves play an important role in determining the path of the jet stream, which travels within the transitional zone between the tropopause and the Ferrel cell. By acting as a heat sink, the Polar cell also balances the Hadley cell in the Earth's energy equation.

The Hadley cell and the Polar cell are similar in that they are thermally direct; in other words, they exist as a direct consequence of surface temperatures; their thermal characteristics override the effects of weather in their domain. The sheer volume of energy that the Hadley cell transports, and the depth of the heat sink that is the Polar cell, ensures that the effects of transient weather phenomena are not only not felt by the system as a whole, but — except under unusual circumstances — are not even permitted to form. The endless chain of passing highs and lows which is part of everyday life for mid-latitude dwellers is unknown above the 60th and below the 30th parallels. There are some notable exceptions to this rule. In Europe, unstable weather extends to at least 70° north.

These atmospheric features are also stable, so even though they may strengthen or weaken regionally or over time, they do not vanish entirely. The Polar cell, Orography and Katabatic winds in Antarctica, can create very cold conditions at the surface, for instance the coldest temperature recorded on the Earth is –89.2 °C at Vostok Station in Antarctica, measured in 1983. Here we must remember two facts, namely: the area gains energy during the six months day and looses energy in the six months night associated with the Sun's North to South and vice-versa travel in a year. Also, the circumpolar vortex is formed more often in the southern cell and less frequent in northern cell and thus makes the southern cell is colder than northern cell. The circumpolar vortex will virtually stop the mid-latitude warm air entering the Polar cell zone. Also net energy in the Southern Hemisphere is lesser than the Northern Hemisphere with more area is covered by oceans. Also, the Sun's energy reaching the Earth is variable with about 11, 22, 44, etc year sunspot cycles. All these play vital role in intra-seasonal and intra-annual weather variations along with long-term temperature variations, more particularly ocean temperatures. Accordingly the middle latitude cell is highly variable.

Some air rising at the polar fronts diverge at high altitude towards the poles to create the polar cell, while the rest moves in the opposite direction to the high level zones of convergence and subsidence at the subtropical ridges on each side of the equator. These mid-latitude counter-circulations create the Ferrel cells that encircle the globe in the Northern and Southern Hemispheres. The Ferrel cell is therefore a secondary circulation feature, dependent for its existence upon the Hadley cell and the Polar cell. It behaves much as an atmospheric ball bearing between the Hadley cell and the Polar cell, and comes about as a result of the eddy circulations (the high- and low-pressure areas) of the mid-latitudes. For this reason it is sometimes known as the "zone of mixing". At its southern extent (in the Northern hemisphere), it overrides the Hadley cell, and at its northern extent, it overrides the Polar cell. Just as the Trade Winds can be found below the Hadley cell, the westerly can be found beneath the Ferrel cell. Thus, strong high-pressure areas which divert the prevailing Westerlies, such as a Siberian high (which could be considered an extension of the Arctic high), could be said to override the Ferrel cell, making it discontinuous.

While the Hadley and Polar cells are truly closed loops, the Ferrel cell is not, and the telling point is in the Westerlies, which are more formally known as "the Prevailing Westerlies". While the Trade Winds and the Polar Easterlies have nothing over which to prevail, their parent circulation cells having taken care of any competition they might have to face, the Westerlies are at the mercy of passing weather systems. While upper-level winds are essentially Westerly, surface winds can vary sharply and abruptly in direction. A low moving pole wards or a high moving equator wards maintains or even accelerates a Westerly flow; the local passage of a cold front may change that in a matter of minutes, and frequently does. A strong high moving pole wards may bring Easterly winds for days.

The base of the Ferrel cell is characterized by the movement of air masses, and the location of these air masses is influenced in part by the location of the jet stream, which acts as a collector for the air carried aloft by surface lows (a look at a weather map will show that surface lows follow the jet stream). The overall movement of surface air is from the 30th latitude to the 60th. However, the upper flow of the Ferrel cell is not well defincd. This is in part because it is intermediatry between the Hadley and Polar cells, with neither a strong heat source nor a strong cold sink to drive convection and, in part, because of the effects on the upper atmosphere of surface eddies, which act as destabilizing influences.

In contrast to the Hadley and Polar systems, the Ferrel system provides an example of a thermally indirect circulation. In the Hadley cell zones in the Northern and the Southern Hemispheres is the zone in which the Sun makes a round in a year and in this cell the day length variations are limited [10 to 14 hours]; Ferrel cell with varying day lengths from 0 to 24 hours and thus associated energy changes are high in this cell. Similarly in Polar cell zones the day is six months and the night is six months.

### 3.2.3 Longitudinal Circulation Features

While the Hadley, Ferrel, and Polar cells are major factors in global heat transport, they do not act alone. Disparities in temperature also drive a set of longitudinal circulation cells, and the overall atmospheric motion is known as the zonal overturning circulation.

Latitudinal circulation is the consequence of the fact that incident solar radiation per unit area is highest at the heat equator, and decreases as the latitude increases, reaching its minimum at the poles. Longitudinal circulation, on the other hand, comes about because water has a higher specific heat capacity than land and thereby absorbs and releases more heat, but the temperature changes less than land. Even at Mesoscales (a horizontal range of 5 to several hundred kilometers), this effect is noticeable; it is what brings the sea breeze, air cooled by the water, ashore in the day, and carries the land breeze, air cooled by contact with the ground, out to sea during the night.

On a larger scale, this effect ceases to be diurnal (daily), and instead is seasonal or even decadal in its effects. Warm air rises over the equatorial, continental, and western Pacific Ocean regions, flows eastward or westward, depending on its location, when it reaches the Tropopause, and subsides in the Atlantic and Indian Oceans, and in the eastern Pacific.

The Pacific Ocean cell plays a particularly important role in the Earth's weather. This entirely ocean-based cell comes about as the result of a marked difference in the surface temperatures of the western and eastern Pacific. Under ordinary circumstances, the western Pacific waters are warm and the eastern waters are cool. The process begins when strong convective activity over equatorial East Asia and subsiding cool air off South America's west coast creates a wind pattern which pushes Pacific water westward and piles it up in the western Pacific. (Water levels in the western Pacific are about 60 cm higher than in the eastern Pacific, a difference due entirely to the force of moving air).

**Walker Circulation:** The Pacific cell is of such importance that it has been named the Walker circulation after Sir Gilbert Walker, an early-20th-century director of British observatories in India, who sought a means of predicting when the monsoon winds would fail. While he was never successful in doing so, his work led him to the discovery of a link between periodic pressure variations in the Indian Ocean and the Pacific, which he termed the "Southern Oscillation".

The movement of air in the Walker circulation affects the loops on either side. Under "normal" circumstances, the weather behaves as expected. But every few years, the winters become unusually warm or unusually cold, or the frequency of hurricanes increases or decreases, and the pattern sets in for an indeterminate period. The Walker Cell plays a key role in this and in the El Nino or ENSO [Southern Oscillation]. If convective activity slows in the Western Pacific for some reason (this reason is not currently known), the climate dominoes next to it begin to topple. First, the upper-level westerly winds fail. This cuts off the source of cool subsiding air, and therefore the surface Easterlies cease.

The consequence of this is twofold. In the eastern Pacific, warm water surges in from the west since there is no longer a surface wind to constrain it. This and the corresponding effects of the Southern Oscillation result in long-term unseasonable temperatures and precipitation patterns in North and South America, Australia, and Southeast Africa, and disruption of ocean currents.

Meanwhile, in the Atlantic, high-level, fast-blowing Westerlies which would ordinarily be blocked by the Walker circulation and unable to reach such intensities form. These winds tear apart the tops of nascent hurricanes and greatly diminish the number, which are able to reach full strength.

**El Niño and La Niña [Figure 3.2a, b, c & d]:** are opposite surface temperature anomalies in the Southern Pacific, which heavily influence the weather on a large scale. In the case of El Niño warm water approaches the coasts of South America which results in blocking the upwelling of nutrient-rich deep water. This has serious impacts on the fish populations. However, the temperature rise is highly variable from event to event **[Figure 3.2a]** -- Sea surface temperature departures from the mean with El Nino episodes – Different groups measured the temperature but they show wide differences **[Figure 3.2b]**. **Figures 3.2c & 2d** present respectively the monthly sea surface temperature anomalies of Nino 3.4 Index [strong and moderate El Nino from 1951]; and the Oceanic Nino Index [ONI]. The ONI presents the clear cut peaks associated with El Nino and La Nina. All these clearly show high year to year and season to season variations.

In the La Niña case, the convective cell over the western Pacific strengthens inordinately, resulting in colder than normal winters in North America, and a more robust cyclone season in South-East Asia and Eastern Australia. There is increased upwelling of deep cold ocean waters and more intense uprising of surface air near South America, resulting in increasing numbers of drought occurrences, although it is often argued that fishermen reap benefits from the more nutrient-filled eastern Pacific waters. The neutral part of the cycle – the "normal" component – has been referred to humorously by some as "La Nada", which means "the nothing" in Spanish.

### 3.2.4 Other Features

In addition, several localized circulation patterns exist, which control the local weather events in different seasons. In India in winter and summer the Western Disturbances play an important role in heat and cold waves in terms of area it extends and intensity it reaches over different parts in south and eastern India. Also, topography like Western Ghats and Himalayan Mountains play vital role in Indian monsoon rains. All these form part of Climate System.

## 3.3 Extra-terrestrial Scenario

**Sunspots:** Climate changes can be triggered in two different ways – by internal and external forces. The internal forces include: Changes in a single climate component, for example, an anomalous ocean current; Changes in the interactions between different climate components, for example, between the ocean and atmosphere; compared to these, the external mechanisms at first glance appear to have nothing to do with the climate system. These include: The very slow drift of continents, which moves land masses into different climate zones over millions of years; the changing intensity of radiation emitted by the sun. The radiation energy of the sun fluctuates over time and changes temperatures on the Earth; volcanic eruptions, which inject ash and sulphur compounds into the atmosphere, influence the Earth's radiation budget and thus affect climate.

J. W. King presented a review on "Sun-weather relationship" in 1975, published in Reviews of Space Science [**King, 1975**]. He states that "Many people have suggested in the past that the weather is influenced by the 11- and 22-year sunspot cycles [**Figure 3.3a, b & c**]. I believe that the accumulated evidence is so compelling that it is no longer possible to deny the existence of strong connections between the weather and radiation changes (electromagnetic and/or corpuscular) associated with a whole range of solar phenomena. Even the most Skeptical scientist who investigates the literature thoroughly will be forced to concede that important aspects of lower-atmosphere behavior are associated with solar phenomena ranging from short-lived events such as solar flares, through 27-day solar rotations to the 11-year, 22 year, and even longer solar cycles."

**Figure 3.3a**: Activity cycles 21, 22 & 23 seen in sunspot number index, TSI, 10.7 cm radio flux, and flares index. The vertical scales for each quantity have been adjusted to permit overlapping on the same vertical axis as TSI. Temporal variations of all quantities are tightly locked in phase, but the degree of correlation in amplitudes is variable to some degree. **Figure 3.3b**: Time versus solar latitude diagram of the radial component of the solar magnetic field average over successive solar rotation. The "butterfly" signature of sunspots is clearly visible at low latitudes. Diagram constructed by the solar group at NASA [regularly updated]. **Figure 3.3c**: the sunspot butterfly diagram – this modern version is constructed and regularly updated by solar group at NASA. **Figure 3.4**: Smoothed mean excess annual rainfall in three northern-hemisphere zones.

**King (1975)** report reveals the fact that in different latitude belts the sunspot cycle present opposite patterns. Because of this, in real time data series, particularly rainfall, presents zonal variations [**Figure 3.4**] – in one zone positively correlated and another zone negatively correlated. In such a scenario, if we take mean of different zones, covering a country or countries, it may not provide significant and meaningful results. Same is the case with temperature.

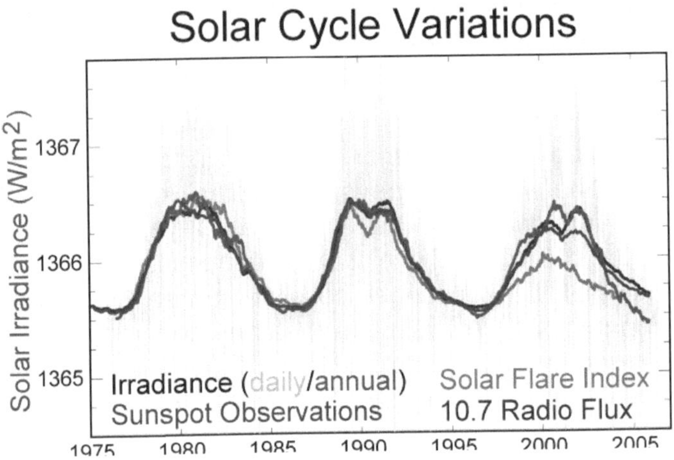


[a]

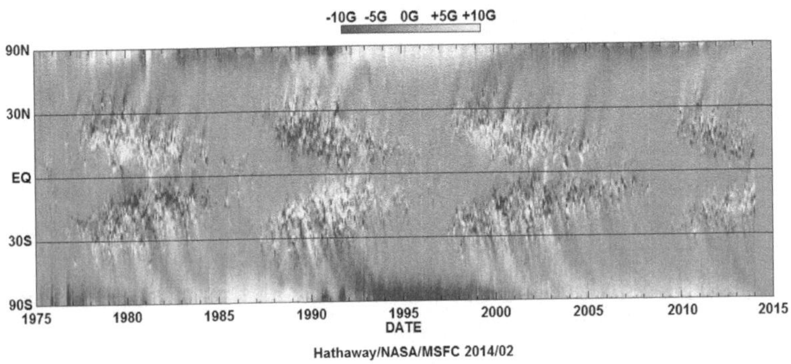


[b]

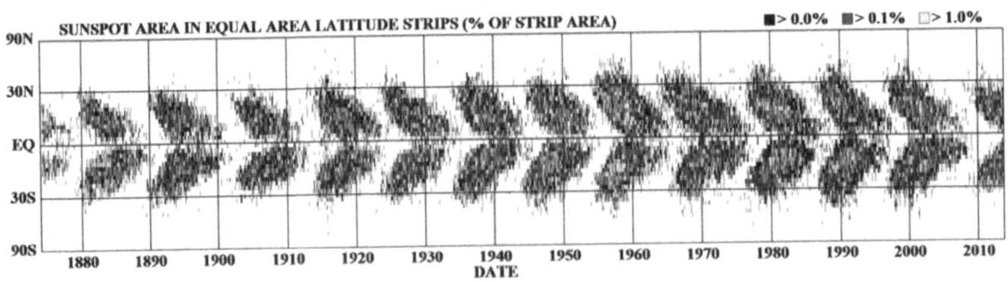


[c]

**Figure 3.3** **[a]** Solar cycle variations, **[b]** Solar latitude diagram, **[c]** Sunspot butterfly diagram

**Solar Radiation:** Global solar radiation and net radiation data series were analyzed through power spectral method to understand whether the data present cyclic variation or not [**Reddy & Lahore, 1977**]. For this purpose used global solar radiation annual estimates for 20 locations of 44 to 67 years and net radiation annual estimates for 8 locations in India well distributed over the country. According to the **WMO [1966]** for power spectrum analysis, the minimum data required is twice to the expected cycle. The data used in this analysis satisfied this condition as expected cycles are 11 and 22 years. The global solar radiation ($R_t$) and net radiation ($R_n$) were derived using the following formulae:

$R_t = aI \times [E_s]^{1/2}$ and $R_n = bI \times T_w$ wherein $I = [1013.2/P]^B \times [N/12.5]$ and $B = 1.5 \log \varphi$

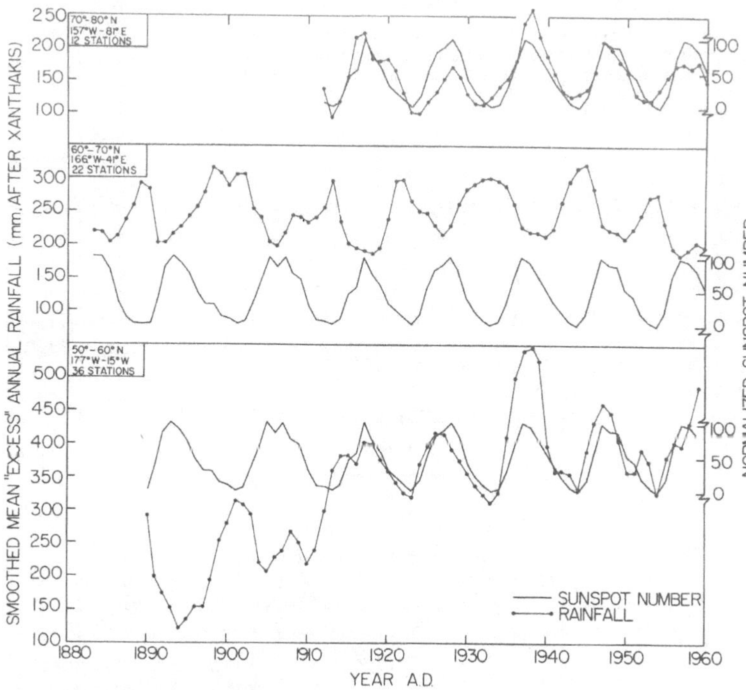


**Figure 3.4** Smoothed mean "excess" annual rainfall in three northern-hemisphere zones versus normalized sunspot numbers

Where 'a' and 'b' are constants vary with seasons given in **Table 3.1**, $E_s$ is the saturation vapour pressure in mbar of mean temperature [(max + min)/2], T is in absolute scale, $T_w$ is the wet bulb temperature [mean of 0830 an 1730 IST observations] in °C, 'l' is the latitude and height correction factor, N is the length of the day in hr, P is the station level pressure [mean of 0830 and 1730 hrs IST observations] in mbar, and φ is the latitude of the place in degrees.

**Table 3.1** Variation of 'a' and 'b' with months – solar radiation model

| Month | a/b | Month | a/b | Month | a/b |
|---|---|---|---|---|---|
| January | 90.0/8.50 | February | 97.0/9.50 | March | 91.0/9.75 |
| April | 85.0/10.63 | May | 77.0/9.75 | June | 65.0/8.50 |
| July | 60.0/8.75 | August | 63.0/9.37 | September | 70.0/9.37 |
| October | 74.0/9.00 | November | 80.8/8.13 | December | 85.0/7.50 |

The following are some of the salient features revealed by the spectral analysis:

$R_n$ – The monthly mean data of unfiltered and filtered using difference filter series showed the annual and semiannual cycles at 99% level; and sub-multiple of sunspot cycle are also seen significant at different levels in unfiltered data [5-6 and 2-3 yr cycles]. The pre-whitened data showed the sunspot cycle and its multiples. The results of power spectrum analysis of annual data confirmed the above results [**Table 3.2**].

The analysis showed persistence in all most all sets of data series. Persistence in the data series means that the power is concentrated in a very narrow band. Such bands can cause errors in the spectral estimates of these frequencies. Therefore, in this study the digital "difference filter" was used to remove low frequency oscillations. Also to understand the low frequency cycles the data series were subjected to low-pass filter, which suppresses the high frequency oscillations.

$R_t$ – The monthly mean data of unfiltered and filtered series showed only the annual and semi-annual cycles significant at 99% level at all the stations. However, the results of pre-whitened data series showed the multiples of sunspot cycle of 10.5±0.5 yr at almost all stations. In order to see clearly the importance of these cycles, the annual mean data series were also subjected to power spectrum analysis. These results are shown in **Table 3.2**. From this table it is seen that under unfiltered data series only at few stations 10.5±0.5 yr cycle was observed and the remaining stations showed 5 to 7 and 2 to 4 yr cycles. The difference filter results showed 6 to 7 years cycle at most stations. The results of low pass filter showed 21.0 yr multiple of 10.5 yr [with two side bands at 14.0 and 8.4]-yr cycle at all most all stations significant at 99% level and at few they are significant at 95% level. Fifty percent of stations showed 42 year cycle also.

**Table 3.2** Results of power spectrum analysis for total solar and net radiation [$R_t$ & $R_n$] – Indian stations

| Station | Unfiltered | Difference filter | Low pass filter |
|---|---|---|---|
| **[a] Total solar radiation [Rt]** | | | |
| Trivandrum | - | 7.0* | 42.0#, 21.0#, 14.0#, 10.5#, 8.4* |
| Port Blair | 6.5* | 6.5* | 14.0#, 10.5#, 8.4# |
| Bangalore | 5.3* | - | 21.0#, 14.0#, 10.5#, 8.4$ |
| Madras | 6.5* | 6.5*, 2.7* | 21.0#, 14.0#, 10.5#, 8.4$ |
| Kodaikanal | 5.3* | - | 21.0#, 14.0#, 10.5#, 8.4$ |
| Hyderabad | - | 6.5* | 42.0#, 21#, 14#, 10.5# |
| Visakhapatnam | 10.5* | 6.0*, 2.2* | 21.0#, 14.0#, 10.5#, 8.4* |
| Bombay | 10.5* | 6.0*, 2.2* | 21.0#, 14.0#, 10.5#, 8.4* |
| Nagpur | 6.0* | 6.0* | 21.0#, 14.0#, 10.5#, 8.4* |
| Calcutta | 3.2* | 3.5* | 42.0#, 14.0#, 10.5#, 8.4# |
| New Delhi | 42.0* | 6.0*, 2.1* | 21.0#, 14.0$, 10.5#, 8.4# |
| Allahabad | 10.5* | 6.0* | 21.0#, 14.0$, 10.5#, 8.4# |
| Jodhpur | 10.5$, 8.4$ | 6.0*, 2.1* | 14.0$, 10.5#, 8.4$ |
| Jaipur | 10.5$, 8.4$ | 6.0*, 2.1* | 14.0$, 10.5#, 8.4$ |
| Srinagar | 10.5$ | 6.0* | 14.0$, 10.5#, 8.4$ |
| Dehra Dun | 10.5$ | - | 14.0$, 10.5#, 8.4$ |
| Sibsagar | 2.5$ | - | 42.0$, 21.0#, 14.0#, 10.5#, 8.4# |
| Rangoon | - | 7.0* | 42.0$, 21.0#, 14.0#, 10.5#, 8.4$ |
| Lahore | 10.5$ | 6.0* | 14.0#, 10.5#, 8.4* |
| Peshawar | 10.5$ | 6.0* | 14.0#, 10.5#, 8.4* |
| **[b] Net radiation intensity [$R_n$]** | | | |
| Madras | 4.3 | 4.0$ | 21.0#, 14.0#, 10.5# |
| Bangalore | 7.0$, 4.7 | 4.7$, 4.0, 3.1 | 21.0#, 14.0#, 10.5#, 8.4# |

*Contd...*

| Station | Unfiltered | Difference filter | Low pass filter |
|---|---|---|---|
| Bombay | 4.3 | 4.0$, 2.4$ | 21.0#, 14.0#, 10.5#, 8.4$ |
| Calcutta | - | 2.8# | 21.0#, 14.0#, 10.5#, 8.4# |
| Rangoon | - | 4.0$ | 21.0#, 14.0#, 10.5# |
| Lahore | 4.7 | 4.7$ | 21.0$, 14.0$, 10.5# |
| Peshawar | 4.0 | 4.0$ | 21.0$, 14.0$, 10.5# |
| Srinagar | 4.0 | 4.0$ | 21.0$, 14.0$, 10.5# |

*Significant at 90% level; $ Significant at 95% level; # Significant at 99% level

**Lunar Atmospheric tides:** It is a common feature with the phase of the moon the ocean/sea tides change. This was linked to gravitational force of the moon. In 40s and 50s studies were carried out to understand the lunar impact on meteorological parameters such as wind and precipitation at short term intervals. **Rao & Reddy [1972a & b]** and **Reddy [1974]** carried out such analyses with rainfall and precipitation at Indian locations. They found positive response but they are of shorter duration. They may be of use in the weather forecasting but not to understand long term changes.

## 3.4 Method of Estimation of Systematic Variations

For the identification of climatic fluctuations, known as short natural rhythmic variation or systematic variations, in met data, we need long term continuous data series. The rainfall data present both systematic and irregular variations over time. Characterization of systematic variations in rainfall data series will help not only in the long-term agricultural & water resources planning but also in long-range weather forecasting [**Reddy, 1993 & 2002**].

**WMO [1966]** presented a procedure for analyzing precipitation data to quantify systematic variations. It is known as power spectrum analysis. In this method sometimes even if the data series has a systematic periodicity, the power spectrum may not represent them. The spectrum analysis throws away all information about the phase of the fluctuations. So, to recover the information on phase other techniques have to be used. The peak refers to a band of wavelengths but not a discrete wavelength. This technique was used by **Reddy [1984]** in the case of Brazilian precipitation data series of 105 locations. Here the main objective was to identify homogeneous zones in terms of their systematic variation patterns over the Northeast Brazil. The results to be reasonably valid the minimum data period required to adopt power spectrum analysis is twice that of expected periodicity or cycle. To get expected periodicity or cycle length, there are other possible alternate methods, such as the moving average technique and iterative regression technique.

The acknowledged purpose of computing moving average technique of climate series is to smooth away the short (rapid) period variations so that the longer (slower) period variations can be revealed more clearly. If there are well defined long period variations in the series this device is quite useful in revealing their form [**WMO, 1966**]. However, this may not specify the quantitative nature of the cycle as such. The 10-year moving average was carried out by **Reddy [1977]** in the case of dates of onset of southwest monsoon over Kerala Coast in India [**Figure 3.5**] & in global average temperature to separate trend from systematic variation, the data was subjected to 10, 30 & 60 year moving average [**Figure 3.6**].

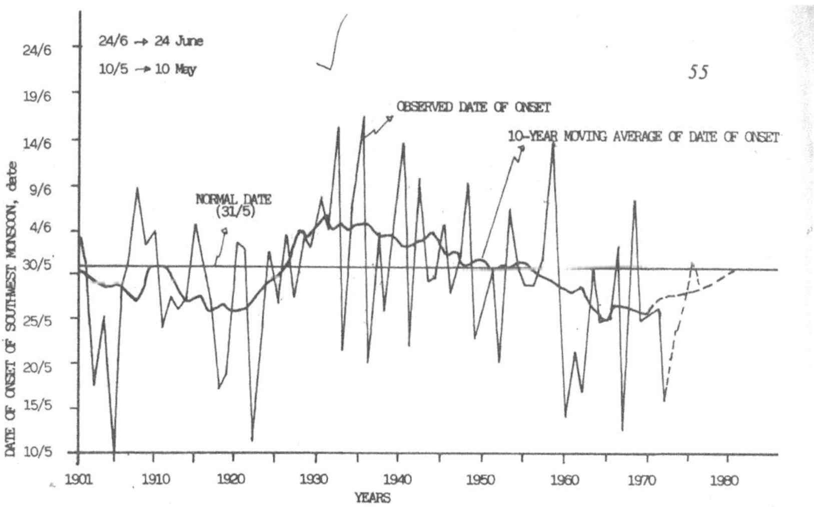


**Figure 3.5** Application of moving average technique to dates of onset over Kerala Coast in India

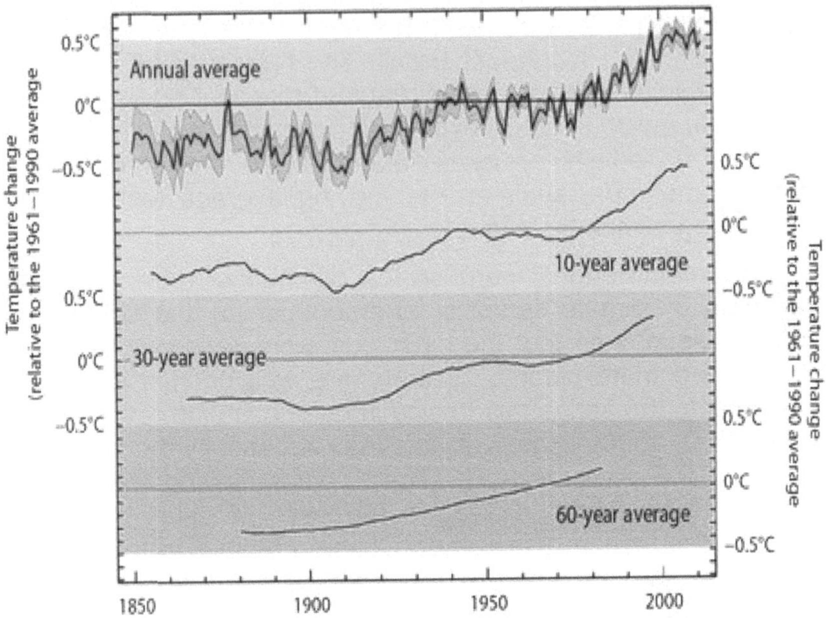


**Figure 3.6** Application of moving average technique to global temperature data series

The computation of raw data and their smoothed spectral estimates in the power spectrum analysis involves computation of auto-correlations and/or auto-co-variances for the defined number of lags. In the case of iterative auto-regression technique, the first step involves the use of these auto-correlations directly to test the significance of different peaks at different lags. The significance test is simple t-test for the appropriate degrees of freedom obtained from the lag number and number of years in the series. Here also, the minimum length of the data series required for this analysis is 2 cycles. The technique was used by **Parthasarathy & Mooley [1978], Reddy et al., [1977], Reddy (1984),** etc.

To be useful for prediction of long-term pattern synthesized curve of the observed periodicities have to be determined. This can be achieved through trigonometric function using amplitude, phase angle and the cycle. The solution of this trigonometric function is achieved through trial and error [iterative] curve matching techniques similar to that of **Reddy, et al. [1984a]** or using maximum likelihood estimates technique. This technique was used by **Reddy (1986)** in the case of Mozambique rainfall series. The same was also used in the case of Ethiopia rainfall series [**Reddy & Mersha, 1990**]. All these are discussed in **Reddy [1993].**

## 3.5 Natural Variations in Precipitation: Case Studies

### 3.5.1 Southern Africa Scenario

Climatic fluctuations in the precipitation data of Mozambique during the period of meteorological records was carried out by **Reddy [1986].** In this analysis precipitation data of 16 locations in Mozambique and five locations from South Africa, Zimbabwe & Malawi were used. As the data period for the majority of the locations of Mozambique are short to apply power spectrum analysis, the periodicities and amplitudes and phase angles were derived using auto-regression approach.

Using the iterative curve matching technique it was observed that none of the locations under study follow the Durban rainfall pattern. The majority of the locations (that were studied) in Zimbabwe follow the rainfall pattern of Catuane in Mozambique. Three locations in Mozambique presented similar to Chileka in Malawi rainfall pattern. Considerable differences are seen in phase angles for these two patterns. **Figure 3.7** presents the pattern of expected year of the commencement of the next below the average annual rainfall patterns.

**Figures 3.8** & **3.9** present the variation of annual rainfall around the mean with years for Catuane in Mozambique and Durban in South Africa, respectively. Through the iterative regression it was found that in Catuane precipitation data series a 54 year cycle and its sub-multiple of 18 year cycle. The same in the case of Durban they are 66 and 22 year cycles. **Table 3.3** presents the respective amplitudes and phase angles of these cycles. Using this data estimated the integrated and projected curves for Catuane and Durban and is presented in **Figures 3.8 & 3.9**. Canberra rainfall series in Australia are similar to those of Durban, around the same latitude belt. **Reddy & Singh [1981]** observed 60-year cycle with sub-multiples of 30, 20 and 10 year cycles in Mahalapye in Botswana precipitation data. This is given in **Figure 3.10.** One important pattern observed in Durban, Mahalapye and Catuane

[**Figures 3.8, 3.9 & 3.10**] with decreasing latitude is the 'M' shape followed by 'W' shape even though the principal cycles respectively presented 66, 60 & 54 years.

**Table 3.3** Cycles, amplitudes and phase angles in Catuane and Durban precipitation data series

| Catuane Cycle [years] | Amplitude [mm] | Phase angle [degrees] | Cycle [years] | Amplitude [mm] | Phase angles [degrees] |
|---|---|---|---|---|---|
| 54 | 2000 | 086.7 | 66 | 250 | 185.5 |
| 18 | 200 | 100.0 | 22 | 350 | 180.0 |

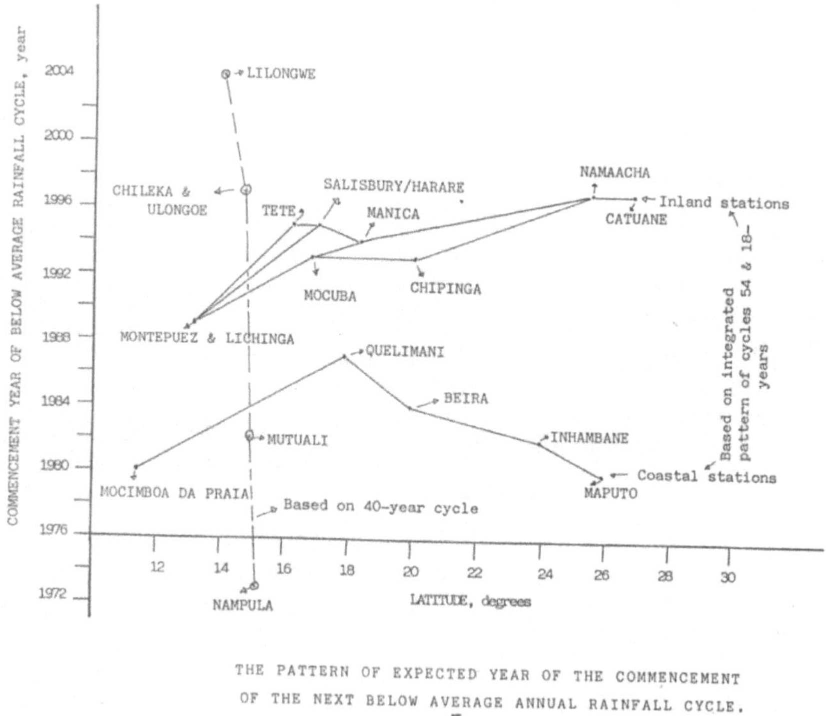


**Figure 3.7** Pattern of expected year of the commencement of the next below average annual rainfall cycle in Mozambique & neighbouring countries

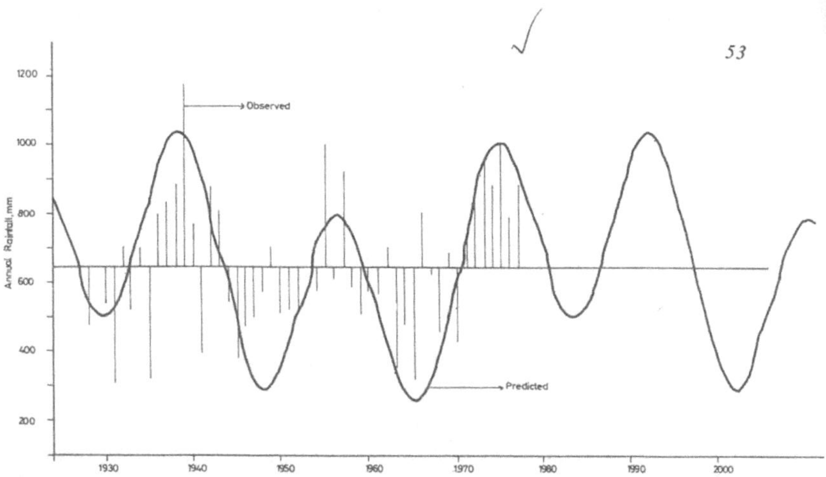


**Figure 3.8** Annual march of annual rainfall (observed and predicted) for Catuane in Mozambique

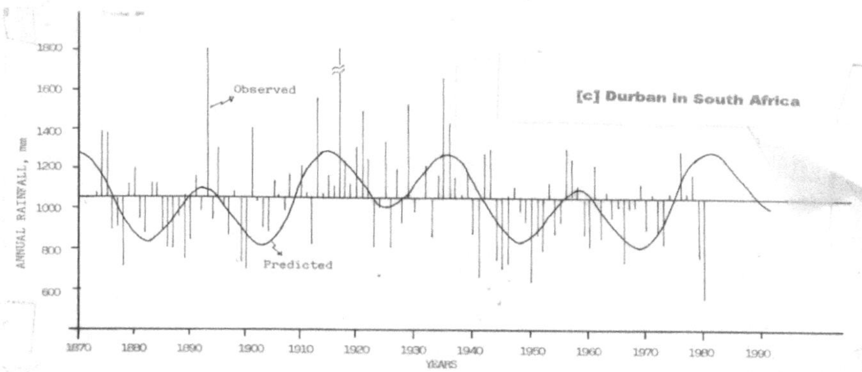


**Figure 3.9** Annual march of annual rainfall (observed and predicted) for Durban in South Africa

## 3.5.2 South America Scenario

In order to assess the cyclic variations in the precipitation data series and to homogenize the precipitation regions in the Northeast Brazil, 70 years annual precipitation data of 105 locations were subjected to power spectrum analysis [**Reddy, 1984**]. In terms of certain cycles the regions above 8° S lat. differs significantly. Also regions between 4° - 8° S lat. differs in some respects from the region below 4° S lat. This is true with respect to coefficient of variation of annual precipitation. The auto-regression of Fortaleza data of 1894-1981 revealed the presence of 52 year cycle and it's submultiples of 26, 13 & 6.5 years. **Figure 3.11** presents the observed and predicted pattern.

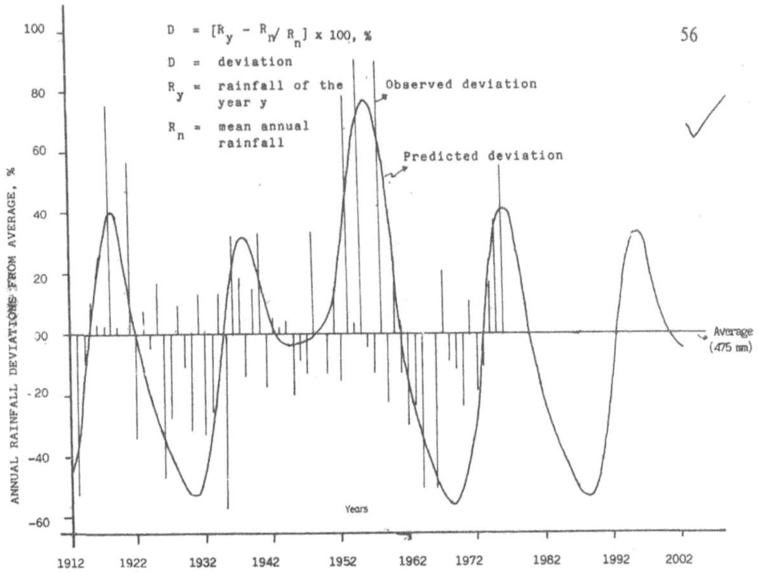


**Figure 3.10** Annual march of annual rainfall (observed and predicted) for Mahalapye in Botswana

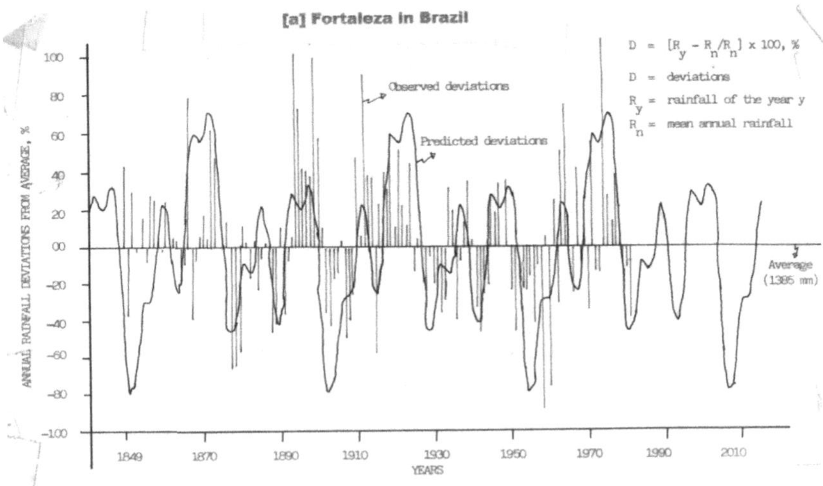


**Figure 3.11** Annual march of annual rainfall (observed and predicted) for Fortaleza in Northeast Brazil

### 3.5.3 Northern Africa Scenario

The precipitation data of 18 stations in Ethiopia were studied to understand the cyclic nature **[Reddy & Mersha, 1990].** As the data presented short period, the data was plotted on graph and fitted to a sine curve of different cycle durations to see best match. The longest cycle of 54-years was observed in Koka of Shea region precipitation while Gonder in Gonder region presented a 52 year cycle. The shortest cycle of 22 years – a sub-multiple of 66 years seen in **Figure 3.9** was observed in Asmara in Eritrea precipitation data. Majority of the areas are covered by 36 and 28 year cycles. Moyale in Sidamo region presented 40 year cycle. Some stations like Addis Ababa and Messawa presented irregular patterns. **Figures 3.12 [a, b, c & d]** present few examples – Gore in a high rainfall zone, Jijiga a bi-model low rainfall zone with sorghum cultivation, Mayola dry-land agriculture zone & Asmara a very low rainfall zone.

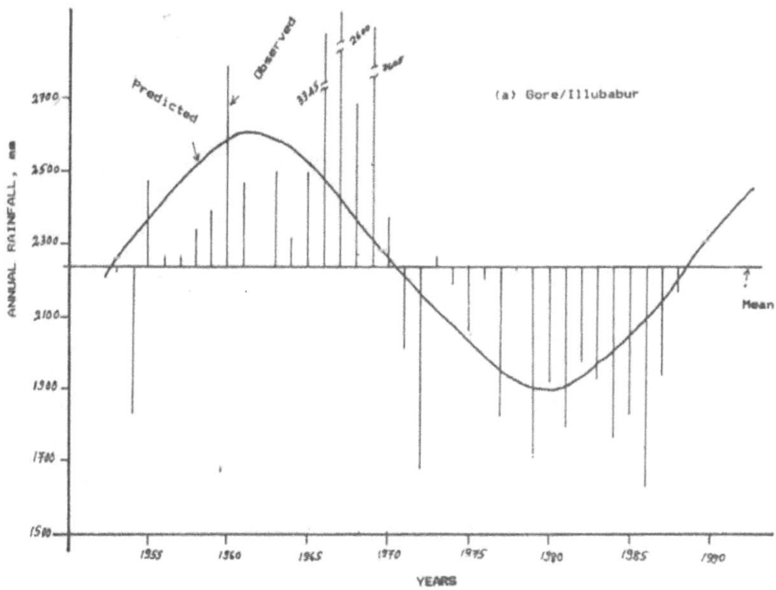

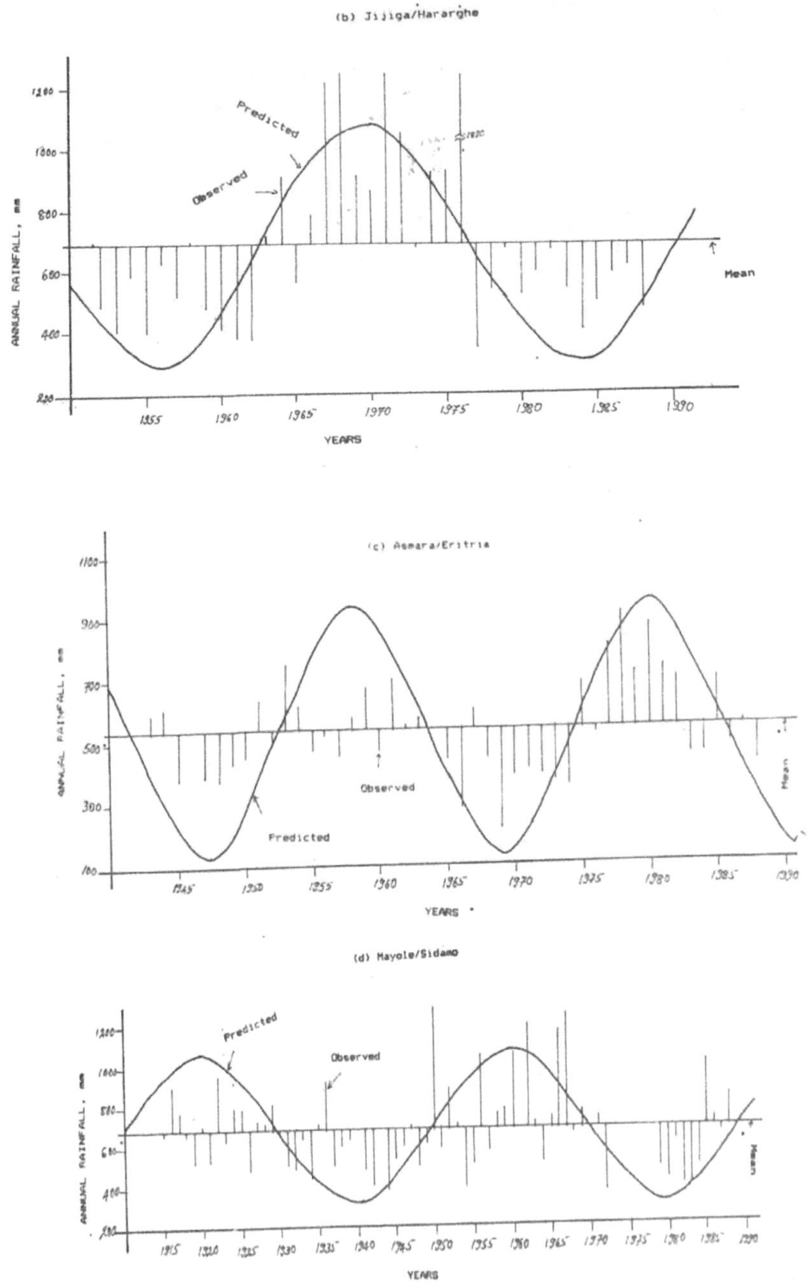


**Figure 3.12** Annual march of annual rainfall (observed and predicted) for four locations in Ethiopia

### 3.5.4 Indian Scenario

**Reddy [2000]** presented some salient features on the Indian Scenario. In 1973, Indian Parliament raised a question on the normal date of onset of Southwest Monsoon over Delhi: Is the normal date correct? To answer this, the task was assigned to I & D section of DDGC of IMD, Pune. I was a member of this group. In this connection I collected the dates of onset and as well dates of withdrawals of Southwest Monsoon over all the 32 meteorological sub-divisions of India [**Reddy, 2002**]. The onset of monsoon starts from Kerala Coast in the State of Kerala, India. I plotted the yearly time series of the dates of onset of the Southwest Monsoon to see whether there is any pattern. Along with the data around the mean date [June 1] presented a 10-year moving average pattern [**Figure 3.5**]. This showed a 52-year cycle [**Reddy, 1977**]. In this figure also presented April wind direction of Stratospheric Winds over Singapore. April wind direction presented a positive correlation with early or late monsoon over June 1.

Indian rainfall is associated with the monsoon system. The word monsoon owes it origin to an Arabic word meaning 'season'. It was used by seamen, several centuries ago, to describe a system of alternating winds over the Arabian Sea. The term is used to connote a seasonal wind, which blows with consistency and regularity during a part of the year, and which is absent or blows from another direction for the rest of the year. The former relates to the winds blowing from the Southwest and termed as the Southwest Monsoon [**SWM**] around June to September. This is also known as Summer Monsoon and Kharif season. The later relates to the winds blowing from the Northeast and is termed as the Northeast Monsoon [**NEM**] around October to December. Major part of the country, including Andhra Pradesh, receives rainfall during SWM. However, Coastal Andhra and Rayalaseema in Andhra Pradesh and coastal Tamil Nadu also receive rainfall during the NEM. IITM [**Parthasarathy et al., 2005**], Pune published a booklet on sub-divisional and country level precipitation data series for months, seasons and annual for the period 1871 to 1994. **Table 3.4** presents the averages and extremes for all India level and three sub-divisions of Andhra Pradesh. 78% of the annual precipitation at all-India level is on an average received during the Southwest Monsoon Season [SWM], i.e., during June to September months. In India 60 year Astrological guide is in practice, wherein it uses to predict the monsoon conditions at the start of New Year around March/April, known as Ugadi in some states [after the Sun crosses the equator on 31st March and moves northwards in India]. The current 60 year cycle started in 1987. Using this, I simply calculated 10-year averages backwards [**Reddy, 2000**]. The plot of these 10-year averages clearly showed completion of two 60-year cycles [**Figure 3.13**]. That means the current 60-year cycle started in 1987 will be ending by 2046. Data from the **Parthasarathy et al. [2005]** publication was used in constructing the figure. The annual march of the all-India Southwest Monsoon precipitation along with the projected curve is given in **Figure 3.14**.

**Table 3.4** Rainfall patterns at all-India and Andhra Pradesh three met sub-divisions

| Region | Parameter | SWM | NEM | Annual |
|---|---|---|---|---|
| All-India | Mean, mm | 852 [78] | 120 [11] | 1090 |
| | C.V., % | 9.9 | 29.0 | 19.5 |
| | The highest, mm | 1020 [120] | 210 | 1348 [124] |
| | The lowest, mm | 604 [71] | 50 | 811 [74] |
| Coastal Andhra | Mean, mm | 507 [42] | 375 [39] | 971 |
| | C.V., % | 22.2 | 38.8 | 19.8 |
| | The highest, mm | 780 [154] | 703 | 1501 [155] |
| | The lowest, mm | 309 [61] | 88 | 532 [55] |
| Rayalaseema | Mean, mm | 422 [60] | 204 [29] | 709 |
| | C.V., % | 28.8 | 41.9 | 21.6 |
| | The highest, mm | 791 [187] | 455 | 1228 [173] |
| | The lowest, mm | 192 [46] | 12 | 226 [32] |
| Telangana | Mean, mm | 722 [80] | 107 [12] | 899 |
| | C.V., % | 23.5 | 60.3 | 21.7 |
| | The highest, mm | 1186 [164] | 310 | 1485 [165] |
| | The lowest, mm | 371 [51] | 02 | 489 [54] |

**Note**: The values in the brackets refer to % of annual value; SWM = Southwest Monsoon, NEM = Northeast Monsoon

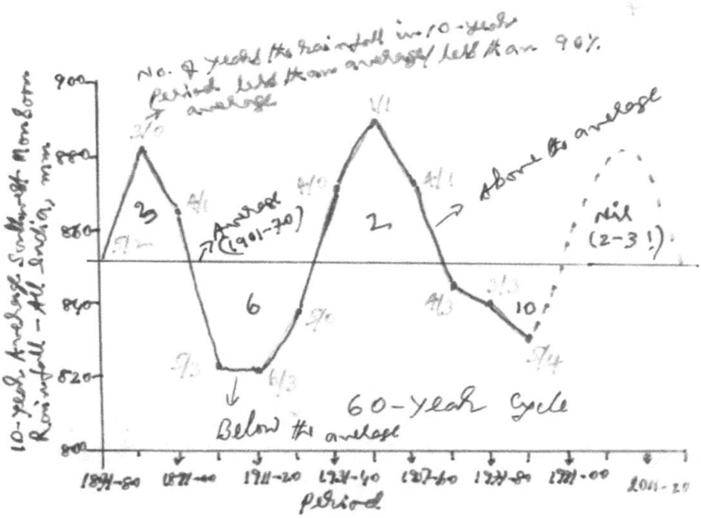


**Figure 3.13** Ten year average march of all-India Southwest Monsoon precipitation

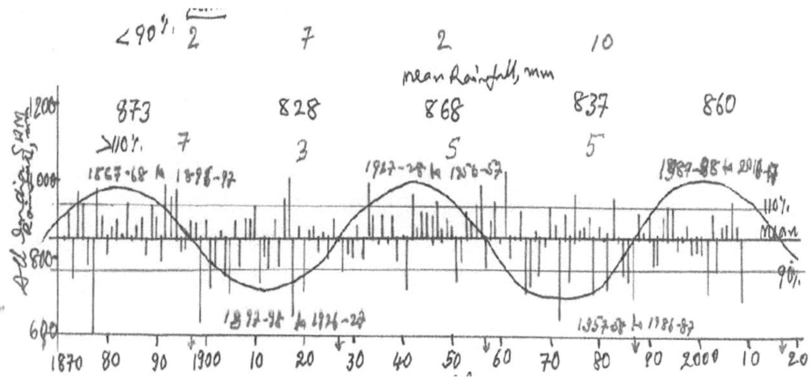


**Figure 3.14** Annual march of all-India Southwest Monsoon rainfall and predicted pattern

### 3.5.5 Andhra Pradesh Scenario

Andhra Pradesh was divided in to three meteorological sub-divisions, namely Coastal Andhra, Rayalaseema and Telangana [now it is bi-furcated in to a separate state of Telangana]. In the case of SWM rainfall, state as a whole and separately the three regions, Andhra Pradesh Rainfall however do not follow the all India rainfall pattern [**Table 3.4**].

Agroclimatic analysis was carried out using weekly rainfall data series of individual years for 80 locations in India [**Reddy, 1993**]. The agroclimatic parameters include available effective growing season [G] and the time of onset of planting rains [S]. When G is less than 5 weeks, it is taken as drought prone year. Based on this, **Figure 3.15** was drawn and that presents drought prone zones. To understand whether the drought proneness presented a systematic variation or not the annual march of the data of G and S for Kurnool was given in **Figure 3.16**. This shows a 56 year cyclic pattern. The drought map was built based on the mean available growing season [one of the parameter used in **Figure 3.16**. For Kurnool the drought condition is 45% of the years on an average. However, during the below the average and above the average 28 year periods they are 70% and 30% years. They play important role in agriculture. Also, the other parameter the commencement of planting rains followed the same around the mean. They are showed in **Figure 3.16**.

**Reddy [2000]** studied the Andhra Pradesh Rainfall. The annual march of SWM and NEM rainfall for the Coastal Andhra sub-division is given in **Figure 3.17**. It presents reverse mirror images for the two seasons – though the magnitude in mm differs. Same will be repeated now, since 2001. Using this analogy, the rainfall of the three meteorological sub-divisions of Andhra Pradesh both SWM and NEM was looked in to. In the SWM rainfall the second cycle of 56 years ended by around 2000; and the third cycle started around 2001. The NEM rainfall followed the same 56 year cycle but in an opposite direction. **Figure 3.18** presents the pattern of 28-year averages for both the SWM & NEM for the three meteorological sub-divisions. As the NEM contribution in Telangana sub-division is low, the NEM cyclic pattern was also weak. These are clearly evident in **Figure 3.19**. However, these patterns showed an increasing trend. Also, one important point to be taken in to account is the pre-monsoon and post-monsoon cyclonic activities, which give copious rains. The frequency of occurrence of cyclonic activity in Bay of Bengal followed the NEM's 56-year cycle [**Figure 3.19**]. To get the in depth reasons for the increasing trend, the annual rainfall was analyzed. This showed a 132 year cycle with the first 66 years being in the below the average part of the cycle with 24 drought years and 12 flood years; and followed this 66 years being in the above the average part of the cycle with 12 drought years and 24 flood years. This cycle ended by 2000 and the new cycle started by 2001 [**Figure 3.20**].

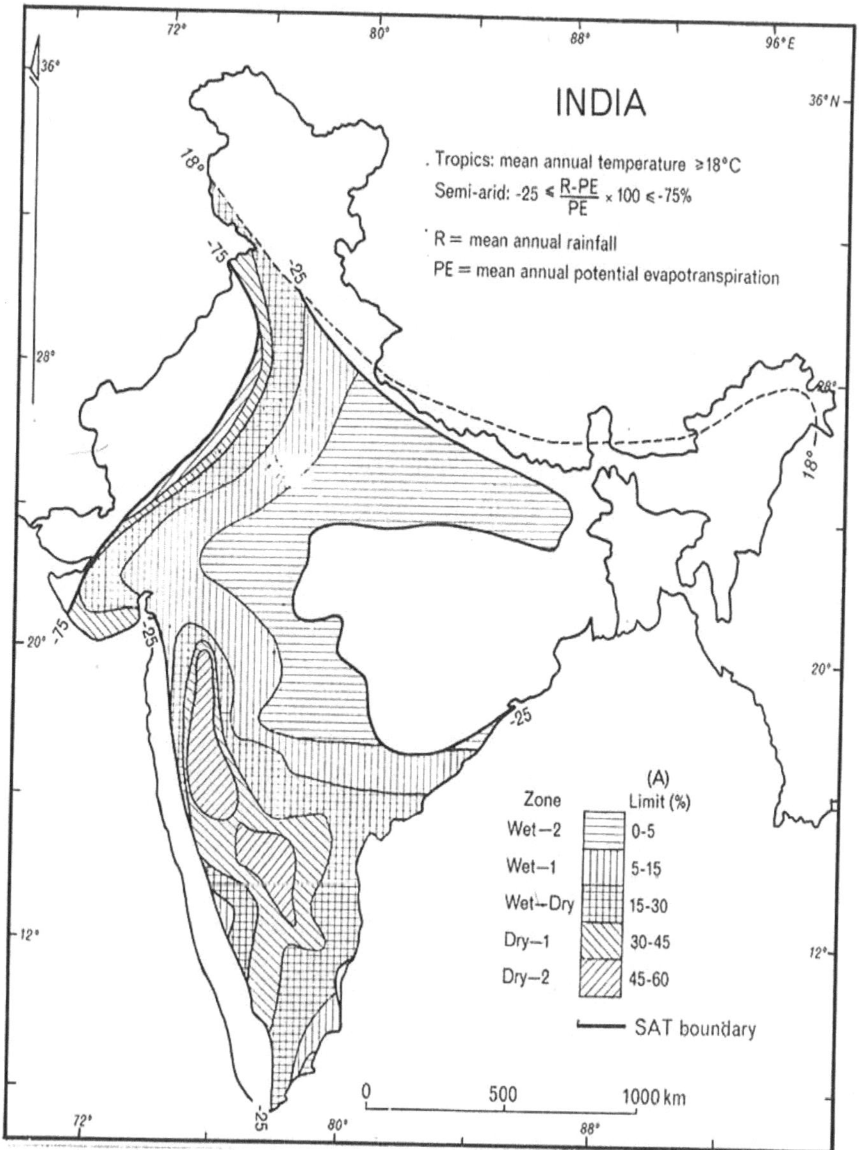


Spatial distribution of the percentage crop failure years or percentage years with G≤ 5 weeks |A|

**Figure 3.15** Drought proneness map of India

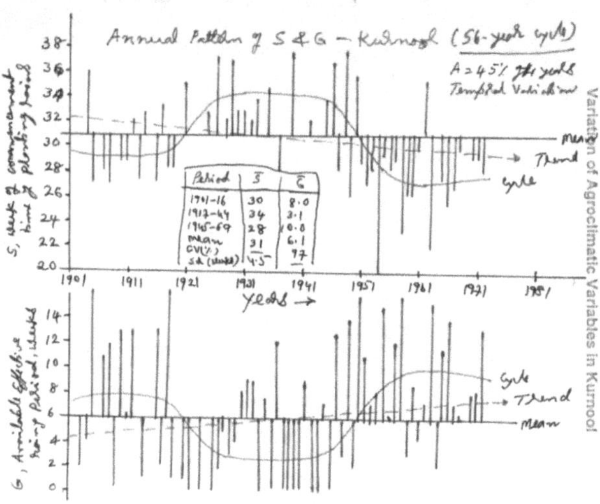


**Figure 3.16** Annual march of S & G for Kurnool pattern and predicted

*(a) Coastal Andhra*

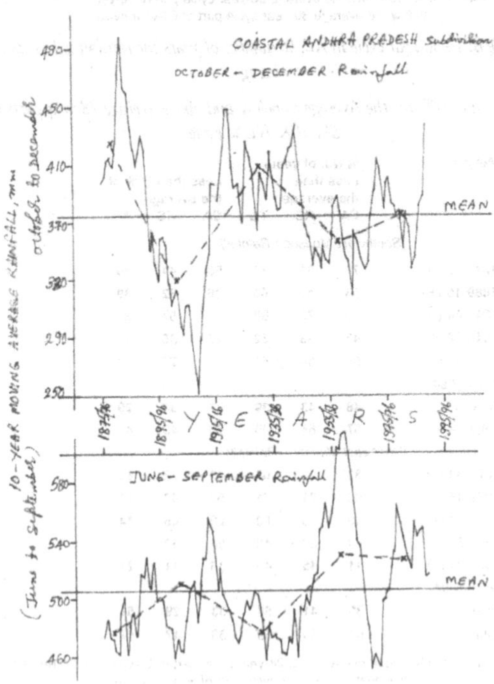


**Figure 3.17** Annual march of SWM & NEM rainfall in Coastal Andhra sub-division – dotted line joins 28-year averages

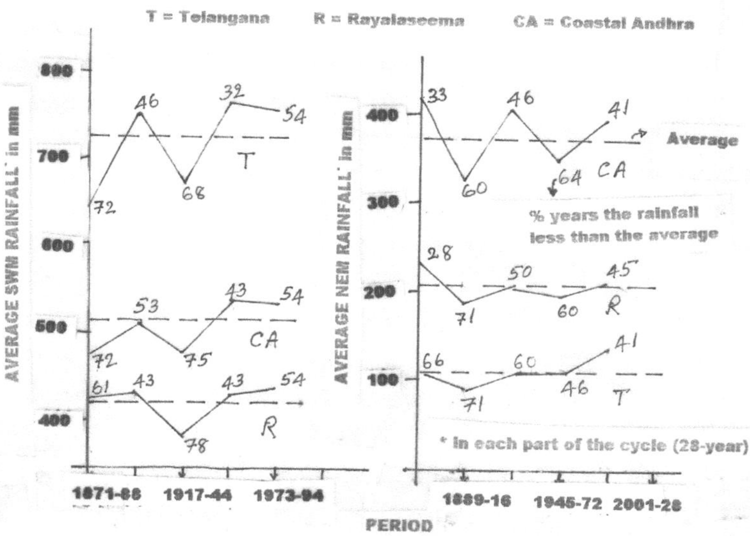


**Figure 3.18** March of 28-year average precipitation at three met sub-divisions and SWM & NEM in Andhra Pradesh

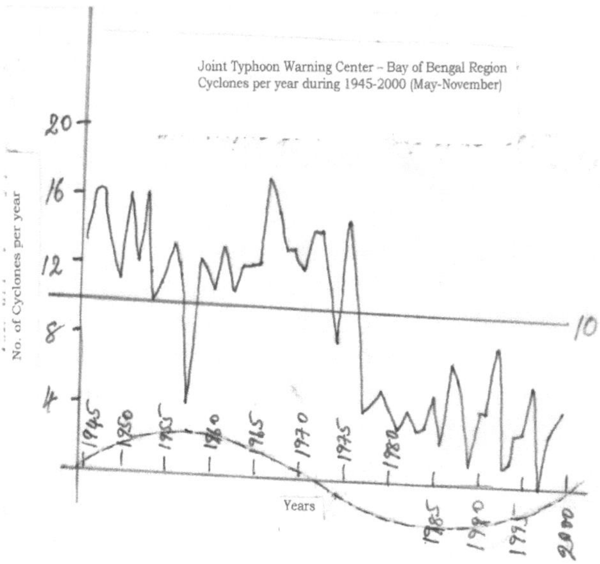


**Figure 3.19** Annual march of cyclonic activity in Bay of Bengal and fitted curve

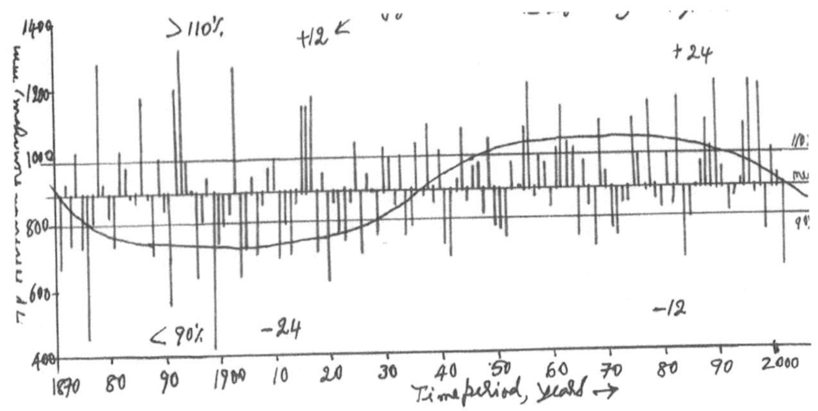

Figure 3.20 Annual march of Andhra Pradesh annual precipitation along with the predicted pattern

## 3.6 General Circulation, Extra-terrestrial & Systematic Variations: A Synthesis

We have seen that the variations seen in the extra-terrestrial phenomenon are also conspicuously observed in general circulation patterns as well in the natural variability in rainfall over different parts of the globe. They are highly region specific in terms of latitude and longitude. Unless these are clearly understood, developing countries are at great risk in coming decades in terms of water resources and agriculture.

In the case of India, the rainfall is drastically modified by orography such as Western Ghats and Himalayan Mountains. Himalayan mountains present a box and thus give copious rains. On the Western side of Western Ghats [wind ward direction] receive copious rains and on the Eastern side of Western Ghats [Lee ward direction] receive poor rains. This is known as shadow zone or drought prone zone. In the shado zone, with the cyclonic activity receive good rains.

We have seen above that the Earth is divided latitudinally in to three cells with quite different circulation patterns. The first cell [low latitude] is apparently driven by the Sun's north to south and vice versa movement in a year. Here the energy variation is not much and through out the year sufficient energy is received. **Reddy & Rao [1976] & Reddy [2008]** presented solar & net radiation and evaporation distribution over India during different months &

seasons and annual. The radiation maps clearly show the latitudinal and longitudinal impact on solar radiation received during different months. When we move to polar cells, the situation is quite different as here energy is received in six months with 24 hours day and the next six months receive no energy with 24 hours night. So, in the first part energy is gained and in the second part energy is lost. Because of this the ice melt show a sinusoidal march with seasons. The maximum ice load is less variable and minimum ice load is highly variable. Also, these variations are within the twice the standard deviation around the mean. The Sun's energy show a marked variation with sunspot cycle -- 11-year and its multiples, namely 22 & 44 years, etc. Year to year variation is controlled by these cyclic changes in energy. They also show changes in latitudinally. All these variations are resulted in the general circulation patterns over different parts of the globe. The middle latitude cell present high variations as it is being a trasition zone between polara and equitorial cells related to Sun's energy quite markedly.

Thus, the integrated system of these three systems play major role in assessing the water resources availablity and success and failure of agriculture on one side and extreme weater events on the otherside..

## 3.7 Concluding Remarks

Where natural variability is present, truncated data will yield highly misleading results. But unfortunately, the media will give high publicity to such studies. The extra-terrestrial phenomenon and as well surface met data present natural variability but they are highly variable with latitudinal zones and as well longitudinal zones. Same is the case with the general circulation patterns.

The average for the globe lead to misleading inferences and based on such information presenting impacts on nature has no meaning as these primarily represent positive and negative changes over different zones. They should be interpreted for the respective local and regional conditions.

To understand and provide long term planning in the water resources and agriculture, analysis of precipitation data at local level is essential and based on such studies identification of homogeneous zones play an important role [**Reddy, 1993 & 2002**].

As long as we neglect these natural variations we are in deep trouble. These define the success and failure of dams and inter-linking of rivers.

All these suggest the fact that blind analysis of data may not yield the correct solution. They must be thoroughly examined like that what was done for Andhra Pradesh three met sub-divisional rainfall data.

Natural variations are not only seen in precipitation data series but also they are seen in sea temperature, polar ice & associated sea levels over different parts of the global oceans. Similarly, they are seen in hurricanes, typhoons, cyclonic activity [**Reddy. 2008**]. Some of these are discussed in **Chapter 2.**

**Current Ethiopian Drought – Is it due to El Nino or a part of Natural Precipitation Cycles?:** Centre for Science and Environment [CSE] a New Delhi based NGO in its' on line Edition of Down To Earth magazine presented a story released on 15th January 2016 by FAO Representative in Ethiopia Anadou Allahoury. The Title of the article is "El Nino: Ethiopia faces worst drought in thirty years". The crop production in Ethiopia dropped by 50 to 90% in some regions and failed completely in the eastern part of the country. The outlook for 2016 is very grim. According to FAO Resident Representative "After two consecutive seasons of failed crops, the success of the main cropping season that starts now will be critical to preventing conditions from worsening". The prevailing El Nino phenomenon is caused crop failures, reduced livestock and pushed around 10.2 million people in Ethiopia in to food insecurity.

Unfortunately, it has become a ritual, to attribute every weather event to El Nino or Global Warming without looking in to weather & climate data of the regions. In 2014, WMO Secretary General in his Word Meteorological Day release attributed the 2013 drought & warm conditions prevailing in Southern Hemisphere to global warming; and the current Ethiopian drought conditions are attributed to El Nino by FAO Representative in Ethiopia. I sent my response to WMO Secretary General saying that the drought conditions and associated warming was part of the natural cycle in precipitation [based on my studies in Brazil, Southern African countries]. Now, the current drought conditions prevailing in Ethiopia are also due to natural cycles in precipitation data series. All this was published in 1980s.

The Ethiopia's case, I as a Team Leader/Chief Technical Advisor of UNDP/WMO project [ETH/86/021-WMO/UNDP] in Addis Ababa/Ethiopia brought a report "Climate fluctuations in the precipitation data of Ethiopia during the period of meteorological record", Agroclimatology Series-4, on 1 June 1990, Addis Ababa, Ethiopia. The results of this study on climatic fluctuations in the precipitation data of Ethiopia suggests that at present in major part of the country is under below the average precipitation pattern. The predominant cycles are 22, 28, 36, etc. For example

- In areas with 28 year cycle [Jijiga, Wonji, Debre-Zeit, Bahir Dar -- low rainfall with bio-model rainfall with dry crops like Sorghum/Pearl Millets], new cycle starts around 1991. That means, around 2006 the below the average 14 year period started.
- In the case of areas with 36 years cycle [Gore, Debre Markes, Gambella, Jimma, Welkite, Combolcha, Dire-Dawa – better rainfall with Coffee, Barley, Wheat] the new cycle starts around 1989. That means, around 2008 the below the average 18 year period started.
- In areas with 22 year cycle [Asmara now in Eritrea – low rainfall], the new cycle starts around 1997. That means, around 2009 the below the average part started.
- Addis Ababa and Massawa [moderate rainfall with the staple food crop Tef] presented irregular variations being bordered to differing cycles with Nagella the below average 18 year period starts around 2019.
- Gonder presents a 52 year cycle with below the average part commencing in 2015 and will continue for 26 years with 6 years better rainfall in the middle: 10 year – 6 years – 10 years.

These years are given in my publication in a table [figures are presented for individual stations]. So, they are part of natural cycle and not due to El Nino. I visited all these regions.

Chapter

# 4

# Impacts of Climate Change: Ground Realities

## 4.1 Introduction

After Copenhagen Summit fiasco in December 2009, significant changes have been taken place in IPCC's concept of global warming and climate change. Some of the conclusions presented in AR4 of 2007 were withdrawn. Al Gore also followed the footsteps of IPCC. But unfortunately both did these only after receiving Noble Prize. However, groups following sensationalization of global warming did not stop from their rhetoric on global warming and climate change. UN agencies also followed this rhetoric. In this connection, an open Letter was submitted to the Secretary-General of the United Nations [125 concerned scientists of the world signed the submission] on 29 November, 2012 – I am one of the signatories --, given as follows:

"Mr. Secretary-General, on November 9 this year you told the General Assembly: "Extreme weather due to climate change is the new normal ... Our challenge remains, clear and urgent: to reduce greenhouse gas emissions, to strengthen adaptation to ... even larger climate shocks ... and to reach a legally binding climate agreement by 2015 ... This should be one of the main lessons of Hurricane Sandy." On November 13 you said at Yale: "The science is clear; we should waste no more time on that debate."

The following day, in Al Gore's "Dirty Weather" Webcast, you spoke of "more severe storms, harsher droughts, greater floods", concluding: "Two weeks ago, Hurricane Sandy struck the eastern seaboard of the United States. A nation saw the reality of climate change. The recovery will cost tens of billions of dollars. The cost of inaction will be even higher. We must reduce our dependence on carbon emissions."

We the undersigned, qualified in climate-related matters, wish to state that current scientific knowledge does not substantiate your assertions.

The U.K. Met Office recently released data showing that there has been no statistically significant global warming for almost 16 years. During this period, according to the U.S. National Oceanic and Atmospheric Administration (NOAA), carbon dioxide ($CO_2$) concentrations rose by nearly 9% to now constitute 0.039% of the atmosphere. Global warming that has not occurred cannot have caused the extreme weather of the past few years. Whether, when and how atmospheric warming will resume is unknown. The science is unclear. Some scientists point out that near-term natural cooling, linked to variations in solar output, is also a distinct possibility.

The "even larger climate shocks" you have mentioned would be worse if the world cooled than if it warmed. Climate changes naturally all the time, sometimes dramatically. The hypothesis that our emissions of $CO_2$ have caused or will cause dangerous warming is not supported by the evidence.

The incidence and severity of extreme weather has not increased. There is little evidence that dangerous weather-related events will occur more often in the future. The U.N.'s own Intergovernmental Panel on Climate Change says in its Special Report on Extreme Weather (2012) that there is "an absence of an attributable climate change signal" in trends in extreme weather losses to date. The funds currently dedicated to trying to stop extreme weather should therefore be diverted to strengthening our infrastructure so as to be able to withstand these inevitable, natural events, and to helping communities rebuild after natural catastrophes such as tropical storm Sandy. There is no sound reason for the costly, restrictive public policy decisions proposed at the U.N. climate conference in Qatar. Rigorous analysis of unbiased observational data does not support the projections of future global warming predicted by computer models now proven to exaggerate warming and its effects.

The NOAA "State of the Climate in 2008" report asserted that 15 years or more without any statistically-significant warming would indicate a discrepancy between observation and prediction. Sixteen years without warming have therefore now proven that the models are wrong by their creators' own criterion.

Based upon these considerations, we ask that you desist from exploiting the misery of the families of those who lost their lives or properties in tropical storm Sandy by making unsupportable claims that human influences caused that storm. They did not. We also ask that you acknowledge that policy actions by the U.N. or by the signatory nations to the UNFCCC that aim to reduce $CO_2$ emissions are unlikely to exercise any significant influence on future climate. Climate policies therefore need to focus on preparation for, and adaptation to, all dangerous climatic events however caused. ---"

IPCC released AR5 in three parts during 2013 & 2014. I presented my observations on these three reports which were published [online] by several forums. Few of them are:

http://whatsupwiththat.com/2013/12/07/ipccs-report-on-climate-change-myths-realities/

http://chimalaya.org/2013/10/08/look-climate-change-issues-in-indian-perspective-rather-blindly-following-ipcc/ [same was also published by "Allvoices"]

Letters to the Editor and Response titled "Climate Change and Global Warming" brought out online by the Indian Geophysical Union [IGU] in, 2014 issue.

An article on "IPCC's WG-II AR5 with reference to India, – the issue is not global warming but the issue is climate change & good governance –". The concluding Para states that "Unfortunately, IPCC and other agencies are attributing all ills of the society to global warming, a phenomenon that is insignificant to influence nature. The issues that are raised under water resources and agriculture by IPCC are nothing to do with global warming but they are primarily related to poor governance and human physical impacts on nature. We must adapt to natural variations in climate. This is what our forefathers did. We need to control pollution [air, water, land & food] and reduce the wastage of food and water resources through better management. We must use local technologies that help in this direction."

Global warming relates to temperature while climate refers to meteorological parameters including temperature. When we are dealing with climate change, it means we are dealing with the changes associated with all the meteorological parameters, more particularly temperature & precipitation, at all scales – local, regional & global -- under climate system & existing general circulation patterns. General circulation pattern system takes in to account the wind and humidity factors. We have seen the scenarios relating to temperature anomaly and precipitation related issues in **Chapters 2 & 3**. The studies include both inbuilt natural variability and human induced trend. In this part the impact of both trend and variability on "nature" issues are discussed. While IPCC put forth several theories on the impact of global warming on nature [**Chapter 2**], several international agencies like NGOs, Banks, Scientific organizations, UN organizations, etc., along with IPCC based on their model predictions brought out several reports on the impact of climate change on "nature". Again here, they all refer to climate change as de-facto global warming, which are in reality related to local-regional climate. With this, they are misleading the developing countries.

## 4.2 Perception on Climate Change & its Impacts

### 4.2.1 IPCC's Perceptions

IPCC-AR5 observed "So far there are no widespread observations of changes in flood magnitude and frequency due to anthropogenic climate changes that is due to 'global warming'. Even their projections imply limited evidence. The projections showed flood hazard increases in "some parts" of the globe. This also presented limited evidence. Recent socio-economic losses from flooding have increased mainly due to "greater exposure and vulnerability". IPCC-AR5 observed "The effect of climate change on crop production is evident in several regions of the world with high confidence. All aspects of food security are potentially affected by climate change, including food access, utilization, and price stability with high confidence". IPCC-AR5 observed "Warming trends and increasing temperature extremes have been observed across most of the Asian regions over the past century with high confidence".

Crop production is highly related to systematic variations component of climate change in precipitation; and the other factors such as food access, utilization and price stability are related to several other localized non-weather factors. The factors of temperature are highly related to systematic variations and ecological changes components of climate change in association with general circulation patterns over different parts of the globe, related to changes in climate system. The extremes are more dependent on which part of the systematic variations the data form part.

IPCC-AR5 presented issues relating these are discussed in the three chapters [3, 7 & 21]. There is some confusion in these discussions, basically because IPCC is using Climate Change instead of global warming wherever it is appropriate. If IPCC could have used global warming as separate entity from natural variability and ecological changes this confusion could have been avoided to a maximum extent. In other words, IPCC should have specified the global warming where its effect is specifically related.

Here it is pertinent to mention, President Vladimir Putin believes that "there is no global warming, that this is a fraud to restrain the industrial development of several countries including Russia. — We found that, while climate change does exist, it is cyclical, and the anthropogenic role is very limited". He is correct to the core as everybody is putting emphasis on energy production issues and nobody putting emphasis on energy utilization or efficiency or low energy consuming technologies. The modern innovative technologies, including IT technologies, are high energy intensive activities and with the progression of time this may further increase non-linearly. The developed nations are passing such technologies to developing countries and thus naturally they need more power.

By 2050 not only population will grow but also around 60% of the world's population will be living in urban areas under new lifestyles under new power intensive technologies including IT sector. This is going to increase non-linearly power consumption levels and as well ecological changes component of climate change. The governments & UN agencies must take in to account these issues while talking on reducing greenhouse gases in power production sector but they don't do such things as it directly affect the multinational companies of the west. This is exactly what I was advocating and working during the past few decades.

### 4.2.2 Wikipedia's Perceptions

Wikipedia talks that nearly all scientists agree that "humans" are contributing to the observed "climate change". A study of academic papers concerning global warming, published between 1991 and 2011 and accessible from web of knowledge, found that among those whose abstracts expressed a position on the cause of "global warming" [though started with humans and climate change], 97.2% supported the consensus view that it is manmade [here it was not specified which factor of manmade?].

In an October 2011 paper published in the International Journal of Public Opinion Research, researchers from George Mason University analyzed the results of a survey of 489 American scientists working in academia, government, and industry. Of those surveyed, 97% agreed that "global temperatures" have risen over the past century and 84% agreed that "human-induced greenhouse warming" is now occurring, only 5% disagreeing that "human activity" is a significant cause of "global warming". National science academies have called on world leaders for policies to cut global emissions.

In the scientific literature, there is a strong consensus that global surface temperatures have increased in recent decades and that the trend is caused mainly by human-induced emissions of greenhouse gases. No scientific body of national or international standing disagrees with this view. If we look at what IPCC said, this statement is inaccurate.

For every coin they have head and tail and similarly here also pros and cons groups are there. In addition, the way IPCC is changing its definitions, coining of words with the passing of time, clearly reflect it is not a science but a political satire. IPCC also differs with other UN agencies on the definition of words and use of words. Even in this section, the first sentence says "humans are contributing to observed climate change" and in the next sentence, they used "papers concerning global warming". It appears they are using climate change as defacto global warming. Also in "humans are contributing", they were talking of anthropogenic greenhouse gases only by excluding the other major human-induced variations, namely ecological changes. These are discussed in the earlier Parts. The studies are most ambiguous in nature to confuse the readers.

Wikipedia also talks that the global warming controversy refers to a variety of disputes, substantially more pronounced in the popular media than in the scientific literature, regarding the nature, causes, and consequences of global warming. The disputed issues include the causes of increased global average temperature, especially since the mid-20th century, whether this warming trend is unprecedented or within normal climatic variations, whether humankind has contributed significantly to it, and whether the increase is wholly or partially an artifact of poor measurements. Additional disputes concern estimates of climate sensitivity, predictions of additional warming, and what the consequences of global warming will be. Now Wikipedia is in right direction in raising those issues. Till to date nobody dared to answer them. Why?

From 1990–1997 in the United States, conservative think tanks mobilized to challenge the legitimacy of global warming as a social problem. They challenged the scientific evidence, argued that global warming will have benefits, and asserted that proposed solutions would do more harm than good.

Some people dispute aspects of climate change science. The fact is, nobody disputed the climate change science, but questioned on the veracity of global warming and using climate change as defacto global warming.

Organizations such as the libertarian Conservative Enterprise Institute, conservative commentators, and some companies such as ExxonMobil have challenged IPCC climate change scenarios, funded scientists who disagree with the scientific consensus, and provided their own projections of the economic cost of stricter controls. Some fossil fuel companies have scaled back their efforts in recent years, or even called for policies to reduce global warming. But, if we look at the above section, it is all a fraudulent observations. They themselves were afraid to use the word global warming to specify the emission issue and instead go on using the word climate change and human-induced.

The world public, or at least people in economically advanced regions, became broadly aware of the global warming problem in the late 1980s. Polling groups began to track opinions on the subject, at first mainly in the United States. The longest consistent polling, by Gallup in the US, found relatively small swings of 10% or so from 1998 to 2015 in opinion on the seriousness of global warming, but with increasing polarization between those concerned and those unconcerned. The first major worldwide poll, conducted by Gallup in 2008-2009 in 127 countries, found that some 62% of people worldwide said they knew about global warming. In the advanced countries of North America, Europe and Japan, 90% or more knew about it (97% in the U.S., 99% in Japan); in less developed countries, especially in Africa, fewer than a quarter knew about it, although many had noticed local weather changes. Among those who knew about global warming, there was a wide variation between nations in belief that the warming was a result of human activities.

By 2010, with 111 countries surveyed, Gallup determined that there was a substantial decrease since 2007–08 in the number of Americans and Europeans who viewed global warming as a serious threat. In the US, just a little over half the population (53%) now viewed it as a serious concern for either themselves or their families; this was 10 points below the 2008 poll (63%). Latin America had the biggest rise in concern: 73% said global warming is a serious threat to their families. This global poll also found that people are more likely to attribute global warming to human activities than to natural causes, except in the US where nearly half (47%) of the population attributed global warming to natural causes.

A March–May 2013 survey by Pew Research Centre for the People & the Press polled 39 countries about global threats. According to 54% of those questioned, global warming featured top of the perceived global threats. In a January 2013 survey, Pew found that 69% of Americans say there is solid evidence that the Earth's average temperature has been getting warmer over the past few decades, up six points since November 2011 and 12 points since 2009.

In such surveys all aspects of climate change as defined by **WMO [1966]** should have been put for vote to get clarity on the subject including global warming. Among the people surveyed, in fact none know the fallacies in the global temperature curve and among climate change and global warming; and among human-induced and anthropogenic greenhouse gases. In such a scenario, you get meaningless results of little use or don't reflect the reality on the peoples' perception on global warming or climate change.

In science such qualitative assessments have no meaning to prove a certain aspect of science. This must be proven through scientific proven method of assessment. Surveys are conducted during election time to show so and so party or person is going to win. They differ from survey to survey. They rarely come true.

Therefore, to avoid any ambiguity on the impact of fossil fuel use, it is essential to refer clearly the word global warming in the surveys. In its place if the word climate change is used, it creates unnecessary confusion on the conclusions drawn. Also, instead of anthropogenic greenhouse gases they are using human-induced, which has other factors also. In fact such surveys only mislead the science. Science does not depend upon, how many people are supporting a theory but depends upon the validity of that theory in real terms. In global warming this is not followed and thus to cover up the deficiencies, they are using the word climate change. Developing countries must get better feedback from UN agencies on the real situation of different components of climate change in different meteorological parameters. Otherwise it will lead to environmental destruction and economic losses at government and people level in developing countries. The developing countries must open their eyes to this fact and instead running for share in green fund, look at how to adapt to climate change in practical sense to minimize risk associated with climate change [see **Reddy, 1993**].

### 4.2.3 Hypothetical Perceptions

Based on unproven theories on global warming, particularly unsound models based temperature projections for doubling of carbon dioxide over prior to industrialization, several agencies such as World Bank, CGIAR groups, Oxfam, Asian Development bank and some reputed institutions including UN organizations [IPCC] brought out reports with wide ranging implications. With their money power they spread erroneous results to developing countries which media publicize it in a ritualistic manner. Unfortunately, these groups rarely read even the IPCC reports in their totality. This is dangerous for the development in developing countries. One such recent report was presented by Amit Garg, Vimal Mishra & Hem H. Dholakia [W.P. No. 2015-11-01, IIM, Ahmadabad] titled “Climate Change and India: Adaptation GAP [2015]”:

http://www.iimahd.ernet.in/assets/snippets/workingpaperpdf/10071603592015-11-01.pdf

First and the foremost is being a prestigious institute in India, how such reports are coming out surprised me. In 2009 also one such report on fertilizer subsidy was presented on which I submitted comments to the authors and sent copy to Prime Minister of India that lead to introduction of Aaadhar card and linking subsidies to it by accepting my proposal giving fertilizer subsidy directly to farmers. Coming to the report, I submitted the following observations to the authors [but received no response from them]: In the "beautiful report" the authors are talking on global warming but titled it with climate change. They should have gone through the IPCC AR5 report for the definition of these two words; Climate change relates to both natural systematic variations [cyclic variations or fluctuations] and man-induced variations – trend --- (a) the author of the book studied the natural variability part in precipitation for several countries including India. Presented in a book "Agroclimatic/Agrometeorological Techniques: As applicable to dry-land agriculture in developing countries, as back as 1993 & the review appeared in an international journal in 1994; & (b) In the case of global temperature – it consists of two primary parts, namely [i] associated with greenhouse effect – according to IPCC AR5 this is more than half [a qualitative statement]; [ii] associated with non-greenhouse effect – known as ecological changes [land & water use and Land & water cover changes] – this is less than half; [c] Greenhouse effect component, again, consists of two parts, namely [i] due to anthropogenic green house gases increase in the atmosphere and [ii] natural forcings such as volcanic activity, etc. What is the share of these two parts was not presented even in qualitative terms. The authors have not made any attempt in this direction.

All India Southwest Monsoon rainfall presents a 60-year cycle -- Started new 60-year cycle in1987. The first part is above the average. Prior to 1987 the 30-year period, that is up to 1957, is below the average part -- in this in ten years the rainfall was less than 90%; The validity of the report mainly depends up on the validity of presentations in chapter 2 "Observed and projected climate change in India"; The report observed "India has experienced substantial changes in mean and extreme climate during the period of 1951-2013. Even if we look at 1931-60 IMD normal Red Book you find this type of extreme changes. The basic question is why such variations are occurring? No such attempt was made by the authors of the report. Only experienced meteorologist can give answers to such issues; Here, it is quite surprising to see that they used 2006-2013 data, though they are talking on 1951-2013; – 1951 is the starting year of global warming as per IPCC; The 1951-2013 data were derived at grid level similar to global temperature wherein actual data was available not even 25% of the globe and this is not different in India. The met stations are mainly established in urban areas for the convenience of measurements and the met network is sparse in rural India and thus the data is biased under grid network – only satellite data eliminates this bias; The differential warming over different zones is associated with the changes in ecological changes [land & water use and Land & water cover] over those zones along with natural cyclic variations in precipitation over different parts as they follow general circulation system or climate system -- if we look at Indian minimum temperature series and maximum temperature series the former presents higher increase and later lower increase; Urban-heat-island effect and rural-cold-island effect are the main parts of ecological changes. The temperature data is biased by the over emphasis in the former covering less than 30% of the area and under emphasis in the later case that constitute more than 70% of the area. Thus the temperature is biased upward; during drought years the temperatures goes up, for example in 2009 and 2002, they have gone up by 0.9 and 0.7 °C. After 1951 nearly thirty years the precipitation was below the average [in 60-year cycle], naturally you get higher temperatures. The data was compared with the base of 1971 to 2000. Before 1971 the data covered 20 years and after 2000, 13 years only.

The south eastern parts present a different rainfall pattern as they receive rains during SWM & NEM and from cyclonic activity. SWM & NEM follow 56 year cycles but in opposite patterns. However, annual rainfall presents 132 year cycle. NEM rainfall is highly variable with the time. In the 132 year cycle: in the 66 year below the average period, 24 years are drought years and 12 years are flood years. In the 66 year above the average period, 24 years are flood years and 12 years are drought years. The temperature follows these patterns.

In the case of heat waves and cold waves associated with the Western Disturbances, the east and southward advance depends upon the high pressure belt around Nagpur and cyclonic activity in Bay of Bengal.

These are some of the practical issues that need attention. This report as well other such reports brought out by several international agencies have no practical utility except misleading the national governments.

Nowadays it became a ritualistic to "pro-global warming groups" to associate any weather event that happens is attributed to climate change [they imply this to global warming] without going in to the past history of that area. For example, In January 2016 California has been hit by a series of winter storms. The groups started declaring all of this as being driven by climate change and claiming that severe weather is getting worse due to climate change. But those who are in the know and who have studied the California weather for decades know that this is no extraordinary event nor is it connected any way to climate change it's just a natural pattern this part of ocean cycles that have been going on for millennia.

## 4.3 Impacts of Climate Change on Nature

### 4.3.1 General

**Figure 4.1** presents the annual march of global temperature under the Northern Hemisphere [NH] and the Southern Hemisphere [SH]; and under the Land and the Ocean. There is vast difference between the NH and the SH on the one hand and on the Land and the Ocean on the other. The annual Ocean temperature march more or less followed the SH annual march. With so much variation, still all UN and other international bodies make statements on impact of global temperature on nature. This is a false alarm.

The so-called model projections in global temperature relative to the regional variations, it is clear that there will be no realistic impact on nature by temperature factor. Also, the seasonal - regional temperature variations are far higher over the model predictions associated with global warming. The impacts more relate to the conditions prevailing over those regions. Though we are talking of global anomaly, the reality is not reflected in regional anomalies.

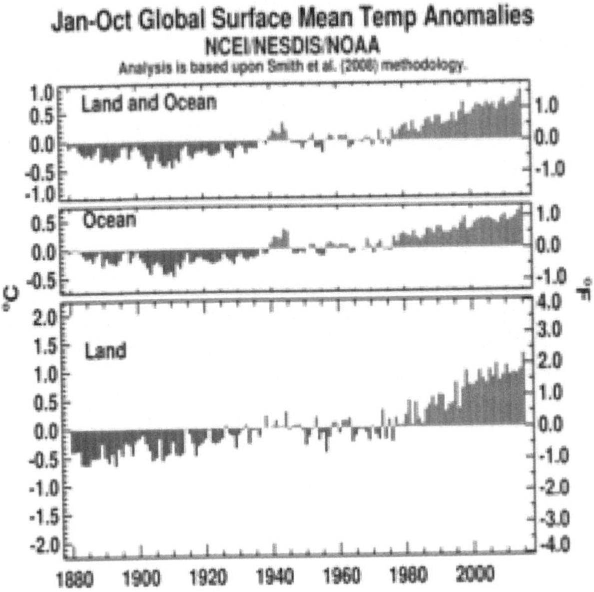


January–October Global Land and Ocean plot

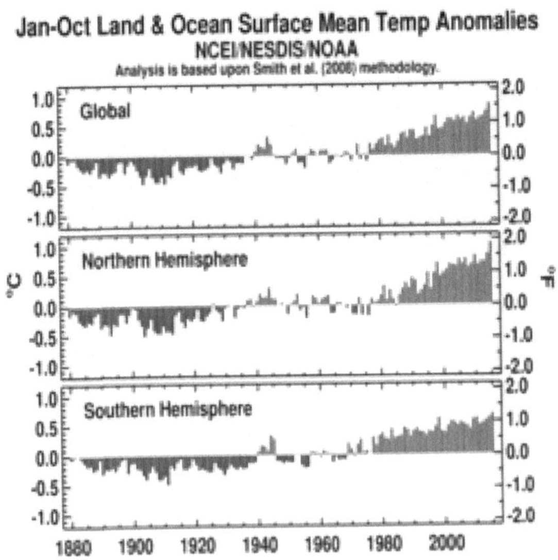


January–October Global Hemisphere plot

The most current data October be accessed via the Global Surface Temperature Anomalies page. January–October Global Land and Ocean plot

**Figure 4.1** Global Temperature Variation for NH & SH and Land & Ocean

## 4.3.2 Weather Systems

I presented my comments on IPCC's AR5, titled "IPCC synthesis Report [AR 5] released on 1/11/2014" -- Myths & Realities. They are presented below:

The following are few headlines taken from the IPCC's Summary Synthesis Report:

1. **Observed changes and their causes:** Human influence on the climate system is clear, and recent anthropogenic emissions of greenhouse gases are the highest in history. Recent climate changes have had widespread impacts on human and natural system.

   1.1 **Observed changes in the climate system:** Warming of the climate system is unequivocal, and since 1950s, many of the observed changes are unprecedented over decades to millennia. The atmosphere and ocean have warmed, the amounts of snow and ice have diminished, and sea level has risen.

   1.2 **Causes of climate change:** Anthropogenic greenhouse gas emissions have increased since the pre-industrial era, driven largely by economic and population growth, and are now higher than ever. This has led to atmospheric concentrations of carbon dioxide, methane and nitrous oxide that are unprecedented in at least the last 800,000 years. Their effects, *together with those of other anthropogenic drivers*, have been detected throughout the climate system and are extremely likely to have been the dominant cause of the observed warming since the mid-20$^{th}$ century.

   1.3 **Impacts of climate change:** In recent decades, changes in climate have caused impacts on natural and human systems on all continents and across the oceans. Impacts are due to observed climate change, *irrespective of its cause*, indicating the sensitivity of natural and human systems to changing climate.

   Evidence of observed climate - change impacts is strongest and most comprehensive for *natural systems*. In many regions, changing precipitation or melting snow and ice are altering hydrological systems, affecting water resources in terms of quantity and quality (medium confidence). Many terrestrial, freshwater, and marine species have shifted their geographic ranges, seasonal activities, migration patterns, abundances, and species interactions in response to ongoing climate change (high confidence). Some impacts on human systems have also been attributed to climate change, with a major or minor contribution of climate change distinguishable from other influences. Assessment of many studies covering a wide range of regions and crops shows that negative impacts of climate change on crop yields have been more common than positive impacts (high confidence). Some impacts of ocean acidification on marine organisms have been attributed to human influence (medium confidence).

1.4 **Extreme events:** Changes in many extreme weather and climate events have been observed since 1950. Some of these changes have been linked to human influences, including a decrease in cold temperature extremes, and an increase in warm temperature extremes, an increase in extreme high sea levels and an increase in the number of heavy precipitation events in a number of regions.

IPCC switched from prediction to projection. The report observed "recent anthropogenic emissions of greenhouse gases are the highest in history" but in global temperature there is a hiatus, plateau, pause or whatever you want to call it for the last 19 years. And yet without taking this in to account IPCC says "warming of climate system is unequivocal". IPCC now brought in new word "Climate System". IPCC was choosy in using the words – most of them are ambiguous like human influence. Nowhere, they referred the word "global warming". IPCC was very tactful this time. They did not use the word around half the global temperature raise as global warming and instead used qualitative words like "dominant". The other qualitative word is "recent". Dominant means there are also other factors contributing to global temperature trend. IPCC is not sure of their % contributions.

In 1.4 IPCC states "number of regions", which means the impact is not uniform to globe. The impact is observed at some zones only. By reading the two sentences under "1 - Observed changes and their causes" one wonders what IPCC wants to tell the world. Here IPCC uses “widespread”. These have no meaning except to confuse and interpret as it suits their ideology. Generally we use the word widespread in weather forecasting referring to a particular region.

The measurements of greenhouse gases started in around 1956 then where from the word “in history” come. When we use a truncated part of rhythmic series “recent” give false alarm lead to misleading conclusions based on the part of rhythmic variation the selected data represents.

Most of the projections are highly hypothetical and not based on facts. IPCC states “it is based on the author teams’ evaluations of underlying scientific understanding and is expressed as a qualitative level of confidence (from very low to very high) and, when possible, probabilistically with a quantified likelihood (from exceptionally to virtually certain), where appropriate, findings are also formulated as statements of fact without using uncertainty qualifiers.” That means they used different qualifiers as per convenience. In science, it should be of uniform in nature.

**Effects of Climate Change on Weather:** The largest chunk of the IPCC synthesis report focuses on the changes in weather patterns and projections related to extreme weather events. It pronounces that these changes are expected to have cascading effects on the health of the economy as well that of the people.

The report observed that the number of monsoon break days is expected to increase while the number of monsoon depressions will see a decline; and at the same time it observed that “mean and extreme precipitation during the Indian summer monsoon is expected to increase”. This cannot be true as with the decrease of depressions rainfall & extreme rainfall decreases as the Indian monsoon precipitation is primarily linked to low pressure systems.

Indian monsoon systems are related to general circulation patterns in different seasons coupled with natural rhythm in meteorological parameters. This we can see from the thumb rules. One such thumb rule is: when low pressure system is located in the Kolkata [Calcutta] region then Hyderabad region will be dry [no precipitation] in southwest monsoon season. This is true even today. Also the onset of monsoon over Kerala Coast follows the pattern of 52 year cycle. The All India southwest monsoon precipitation follows the 60-year cycle. With this, from 2017 to

2046 the precipitation will be below the average part of 60-year cycle. In this period expected more drought years and less flood years and thus frequency of occurrence of floods in NW Indian rivers will be few; and consequently the temperature will be higher and thus evaporation. This will affect the dry-land agriculture in many years.

**Extreme weather events: cyclones, floods and droughts:** The IPCC report observed that the frequency of cyclones likely decrease in 2030s. Floods and droughts are likely to increase in India since there will be a decline in seasonal precipitation, coupled with extreme precipitation during monsoon.

**Cyclonic activity --** During 1891 to 1990 cyclonic disturbances in Arabian Sea were 259 [143 depressions, 47 cyclonic storms & 68 severe cyclonic storms] and in Bay of Bengal they were 1212 [767 depressions, 260 cyclonic storms & 185 severe cyclonic storms]. Majority of cyclonic storms and severe cyclonic storms were seen in the pre-monsoon and in the post monsoon seasons. In the pre-monsoon season the temperature reaches to maximum peak and during post-monsoon season the temperature comes under declining phase to reach minimum in a sinusoidal seasonal temperature curve following the Sun. The data presented by Joint Typhoon Warning Centre on cyclones per year during 1945 to 2000 showed a 56 year cycle. This follows the northeast monsoon precipitation pattern of Andhra Pradesh State in India. During 1945 to 1972 the frequency was above the average and from 1973 to 2000 the frequency was below the average. 2001 to 2056 will repeat 1945 to 2001 pattern. That means 2029 to 1956 follow the frequency of below the average. The average number of cyclones per year was 10; during below the average part they were 2 to 8 and during above the average part they were 10 to 16 [except in two years with 4 & 8].

In 1977 severe cyclone around 10,000 people died, in 1990 cyclone only around 1000 people died; though the later one is more severe than the 1977 cyclone. The death toll was higher in 1977 as the area is low lying – below sea level. In the 2014 severe cyclone the loss of human life was around 3 to 5 persons but latter addition was around 50. So the impact relates to terrain conditions, seawall protection [natural mangroves, etc], destruction of the coast line due to illegal developmental activities and disaster management actions. The cyclonic tracks covered the entire east coast line of Andhra

Pradesh, with more frequent around Srikakulam, East Godavari, Krishna, and Nellore districts. Though it was made possible to bring down human losses with better disaster management, the same was not achieved in animal, crop, property losses. This is part of natural rhythm and nothing to do with global warming. **Table 4.1** presents the precipitation variations in three meteorological sub-divisions in Andhra Pradesh. **Table 4.2** presents the 28 year average in SWM & NEM over the three met sub-divisions of Andhra Pradesh. Both the monsoon rainfall presents 56 year cycles but in opposite direction – see **Chapter 3**.

**Table 4.1** Precipitation variations in three met sub-divisions of Andhra Pradesh

| Met Sub-division | parameter | SWM | NEM |
|---|---|---|---|
| Coastal Andhra | Mean*, mm | 507 (52) | 375 (39) |
| | C.V., % | 22.2 | 38.8 |
| | Lowest | 309 | 088 |
| | Highest 780 | 703 | |
| Rayalaseema | Mean, mm | 422 (60) | 204 (29) |
| | C.V., % | 28.8 | 41.9 |
| | Lowest | 192 | 012 |
| | Highest | 791 | 455 |
| Telangana | Mean, mm | 722 (80) | 107 (12) |
| | C.V., % | 23.5 | 60.3 |
| | Lowest | 371 | 002 |
| | Highest | 1186 | 310 |

* the values in brackets are % of annual mean; SWM = southwest monsoon; NEM = northeast monsoon; C. V. = coefficient of variation – Estimates based on 1871 to 1994 data published by IITM, Pune, India

**Table 4.2** The 28-year averages during SWM & NEM in three met sub-divisions of Andhra Pradesh

| | Period part of cycle 28-year average precipitation, mm | | | | | | |
|---|---|---|---|---|---|---|---|
| | | SWM | | | NEM | | |
| **Period** | **Climate Cycle** | **CA** | **R** | **T** | **CA** | **R** | **T** |
| 1889-16 | above the average | 509 | 432 | 743 | 330 | 188 | 091 |
| 1917-44 | below the average | **477** | **386** | **673** | **410** | **205** | **104** |
| 1945-72 | above the average | 532 | 423 | 761 | 356 | 198 | 108 |

CA = Coastal Andhra, R = Rayalaseema, T = Telangana 28 year averages represent below the average and above the average periods of 56 year cycle – the cycles of SWM and NEM follow opposite patterns

IPCC report states that the India's temperature will increase from 1.7 to 2.2 °C over 1970 level **in 2030**. But, the global average temperature has not shown any increase for the last 19 years [1998 to 2015] – in this report IPCC used term like in 2030 instead by 2030!!! The actual data shows the raise in minimum temperature is higher than that in maximum temperature and yet the value of 2011 is less than 1970 in the case of minimum temperature; the same in the case of maximum temperature shows a small raise from 1970 to 2011. After correcting to 60-year cycle component the trend is very little. The global temperature 60-year cycle varied between - 0.3 to and 0.3 °C [**Figure 1.6a**]. In such a scenario, can we expect that much rise in Indian temperature? Also, in India urbanization and thus urban-heat-island affect is over emphasized with dense met stations confining to urban areas. However, the rural-cold-island affect with sparse met network the impact was not accounted proportionately, that is under emphasized. This results a positive rise. The temperature is highly variable from season to season; year to year; soil to soil; coast to inland to hilly terrain; etc. The temperature follows the opposite pattern with floods and droughts. Also, crop development is influenced not only by temperature but also by the soil moisture, relative humidity, soil temperature, soil fertility, plant density, fertilizer use efficiency, etc.

The raise in minimum temperature is more in association with the urban heat-island effect. Also, the met network is more concentrated in urban areas. Thus it is over emphasized in the averaging of temperature. The met network is sparsely distributed in rural areas and thus it is underemphasized in the averaging of temperature. The net result in less than the observed average temperature. The crops are mainly located in rural areas and thus the temperatures in these zones are below the average.

**Cold waves and heat waves:** In the case of India the heat and cold waves are associated with western disturbances from northwestern region. Their movement to south and east are related to general circulation patterns in the south and east. The extreme temperatures [maximum or minimum] have not exceeded the extremes presented in the 1931-60 normal [IMD – Red Book]. Temperature presents opposite pattern to precipitation pattern. That is during wet period the temperature comes down over the average and during the dry period the temperature goes up over the average. The cyclic part of the temperature is part of observed temperature. When local terrain, forests are destroyed we get changes in meteorological parameters including temperature.

**Health:** Pollution-induced changes in air and water quality, as well as changes in weather pattern, are expected to have wide-reaching effects on the health of Indians, according to the IPCC report.

To avoid the criticism on the impact of global warming on health, they mixed this with pollution. In fact we are fighting the IPCC and UN on this issue for the long. Don't mix the pollution with global warming when we are talking of health hazards on life farms. It is entirely different issue and not related to global warming.

The presentation in Synthesis Report of AR5 is quite different from the previous two AR5 reports. This report is filled with ambiguous statements. To avoid the confusion particularly in relation to the impact aspects instead of using the generalized word climate change, use of specific part of climate change, namely global warming, ecological changes, natural variability, etc gives the clarity to public on the issue of global warming.

Separate the global warming component from ecological change component. Tell the public, is global warming component since 1951 to date is around 0.1 °C or between 0.1 and 0.2 °C? Then only we get the real picture on what will be the global warming by 2100! At present lump sum is used as global warming. The ecological changes part impacts at local and regional scales and not at global scale. As the surface data present an over estimate of ecological change component, use the satellite data for the purpose of separating global warming component from ecological change component.

In the case of sea level change, separate the impact due to global warming from natural variation, land sinking from oil, gas & water extraction, destruction of coast line – particularly protective walls of nature, etc., components.

In the case of changes in ice, separate the impact of global warming from the impacts of natural disasters such as earthquakes, volcanoes, nuclear tests, and sporting, etc., components.

There is no impact of global warming on Indian weather systems, water resources availability and agriculture production. This is even up to 2100. Natural variability impacts are depicted as caused by global warming only. Extreme weather events are part of natural variability.

IPCC at last recognized the importance of pollution on health. It is also an important factor even for agriculture production.

IPCC must not use ambiguous terminology and present where the data is quantitatively validated.

Some reports suggest that human-caused global warming is changing weather systems and making heat waves and droughts more intense and more frequent. The May 2014 National Climate Assessment report said human-caused climate changes, such as increased heat waves and drought, "are visible in every state". A Sep. 2014 American Meteorological Society study found that human-caused climate change "greatly increased" (up to 10 times) the risk for extreme heat waves in 2013. According to an Aug. 2012 study published in the *Proceedings of the National Academy of Sciences*, there is a "high degree of confidence" that the Texas and Oklahoma heat waves and drought of 2011, and heat waves and drought in Moscow in 2010, "were a consequence of global warming" and that "extreme anomalies" in weather are becoming more common as a direct consequence of human-caused climate change. A 2015 study found that globally, 75% of extremely hot days are attributable to warming caused by human activity. Some other reports present exactly other way. See for example:

If we look at **Figure 4.2**, it gives the clear picture of Hurricane season. It is following a normal distribution. **Figure 4.3** presents the occurrence of most extreme temperatures by continent.

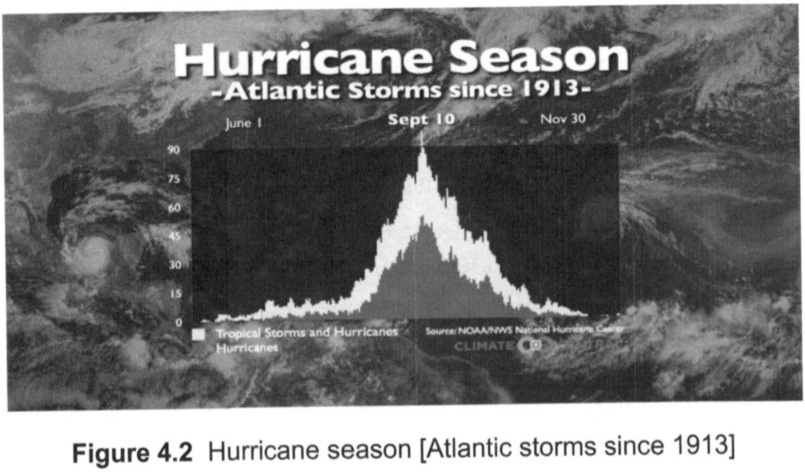


**Figure 4.2** Hurricane season [Atlantic storms since 1913]

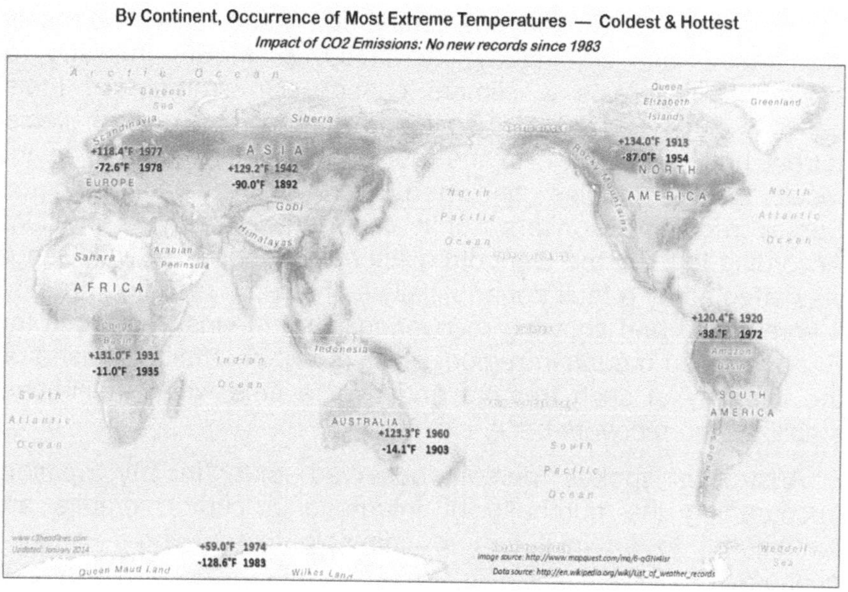


**Figure 4.3** Occurrence of most extreme temperature by continent

In global warming the greenhouse effect changes the energy in to temperature by carbon dioxide. How much is the contribution of this changed temperature to global temperature rise is the basic question. For this mathematical equations were used but IPCC is using trial and error approach and thus the sensitivity factor monotonically reduced from SAR to AR5. The sensitivity factor in AR4 varied between 1.55 to 2.35 with average of 1.95. The same in the case of AR5 varied between [watts per meter square per $^oK$] 1.00 to 2.25 with the average of 1.55 and thus the respective temperatures by 2100 with the doubling of Carbon Dioxide are expected as: 2.2 to 4.4 with 3.0; and 1.7 to 3.9 with 2.2 $^oK$. This shows that there is no way to quantify the global warming component in global temperature rise and because of this IPCC used qualitative statement like more than half in which include other greenhouse forcings.

Environment Minister Prakash Javdekar in a reply to Rajya Sabha informed that of the 2018 Himalayan Glaciers representing climatically diverse terrains in the Himalayas [Karakoram, Himachal, Zanskar, Uttarkhand, Nepal and Sikkim regions] were studied, majority 1752 [86.8%] glaciers remained stable and rest presented some melting and some building. Since last two years including this year Himalayans getting heavy snowfalls. The results are based on joint study by ISRO & MoEF [ministry of environment, forests & climate change, Government of India] between 2004 and 2011. Also found number and rate of glacier retreat had come down in the last decade, as compared to the results of other studies conducted before 2001. Indian Geological Survey regularly monitors some important Himalayan Glaciers. According to them the Gangotri glacier that feeds the River Ganga was affected by a fault zone. Initially it showed a receding but now it is reversed and showing recovering. That means tremors in the Earth in those mountain region, glaciers will be affected based on the intensity of such tremors and over a time when the impact stabilizes the recovers back.

Australian spokes person observed that "In my opinion, encouraging the purchase of international carbon credits are tantamount to an invitation to criminals to loot the Australian economy."

Increased hurricane activity and other extreme weather events are a result of natural weather patterns, not human-caused climate change [**Reddy, 2008**]. According to a 2013 report from the Tropical Meteorology Project at Colorado State University, the increase in human-produced $CO_2$ over the past century has had "little or no significant effect" on global tropical cyclone activity. The report further states that specific hurricanes including Sandy, Ivan, Katrina, Rita, Wilma, etc., are not the direct consequences of human-caused global warming. Between 1995-2015 increased hurricane activity (including Katrina) was recorded, however, according to the NOAA, it was not the result of human-induced climate change; it was the result of cyclical tropical cyclone patterns, driven primarily by natural ocean currents. Many types of recorded extreme weather events over the past half-century have actually become less frequent and less severe. Professor of Earth and Atmospheric Sciences at the Georgia Institute of Technology, Judith Curry, PhD, states that she is "unconvinced by any of the arguments that I have seen that attributes a single extreme weather event, a cluster of extreme weather events, or statistics of extreme weather events" to human-caused climate change. Richard Lindzen, PhD, Emeritus Professor of Meteorology at the Massachusetts Institute of Technology, also states that there is a lack of evidence connecting extreme weather events such as hurricanes, tornadoes, droughts, or floods, to human-caused global warming.

Without looking in to the historical data, reports noted that "Dramatic changes in precipitation, such as heavier storms and less snow, are another sign that humans are causing global climate change. As human-produced greenhouse gases heat the planet, increased humidity (water vapor in the atmosphere) results. Water vapor is itself a greenhouse gas. In a process known as a positive feedback loop, more warming causes more humidity which causes even more warming. Higher humidity levels also cause changes in precipitation. According to a 2013 report published in the *Proceedings of the National Academy of Sciences*, the recorded changes in precipitation over land and oceans "are unlikely to arise purely due to natural climate variability." Higher temperatures from global warming are also causing some mountainous areas to receive rain rather than snow. According to researchers at the Scripps Institution of Oceanography, up to 60% of the changes in river flow, winter air temperature, and snow pack in the western United States (1950-1999) were human-induced. Since 1991, heavy precipitation events have been 30% above the 1901-1960 average in the Northeast, Midwest, and upper Great Plains regions. A 2015 study found that global warming caused by human actions has increased extreme precipitation events by 18% across the globe, and that if temperatures continue to raise an increase of 40% can be expected.

Permafrost is melting at unprecedented rates due to global warming, causing further climate changes. According to a 2013 IPCC report there is "high confidence" (about an 8 out of 10 chance) that anthropogenic global warming is causing permafrost, a subsurface layer of frozen soil, to melt in high-latitude regions and in high-elevation regions. As permafrost melts it releases methane, a greenhouse gas that absorbs 84 times more heat than $CO_2$ for the first 20 years it is in the atmosphere, creating even more global warming in a positive feedback loop. By the end of the 21st century, warming temperatures in the Arctic will cause a 30%-70% decline in permafrost. According to a 2012 report, as human-caused global warming continues, Arctic air temperatures are expected to increase at twice the global rate, increasing the rate of permafrost melt, changing the local hydrology, and impacting critical habitat for native species and migratory birds. According to the 2014 National Climate Assessment, some climate models suggest that near-surface permafrost will be "lost entirely" from large parts of Alaska by the end of the 21st century".

Unfortunately the same is routinely repeated by World Bank, Asian Development Bank, etc., similar to what Harjeet Singh presented in his article. Here let us not go in to details but see the comments made by Harjeet Singh on two important points. They are as follows:

1. In the last three years itself, India confronted enormous devastation caused by the Uttarkhand floods, by Cyclone Phelin, Kashmir floods and an unprecedented heat wave this year

   My observations: These are not unusual to India. They were there in India in the past, in the present and will be there in future. The devastations caused by floods in Uttarkhand and Jammu & Kashmir were man induced. The government closed their eyes while the unscrupulous people encroaching the river banks and constructing buildings. This resulted reduced channels for river water flows and thus they overflowed by washing away the buildings and people along with their property enroot. In the case of cyclonic activity, India reduced human deaths but not animal and property losses. Here proper planning plays an important role. Heat & cold waves are common to India in summer & winter respectively in association with Western disturbances, a general circulation pattern. The author presented an article on this as back as 1978 in India Meteorological Department Journal. This year's heat wave followed the Figure 7a & b pattern – high pressure belt position and associated wind patterns -- in the article only. They are nothing to do with global warming.

   November 2015 floods under the Northeast Monsoon in Nellore District are the result of destruction of natural water flow channels. This resulted cutting of roads by flood waters.

2. At the Paris climate summit this December, 195 countries will come together to script a new agreement to limit the temperature rise to less than 2.0 °C and deal with increasing impacts of climate change.

My observations: Paris summit is for collecting 100 billion dollars under the disguise of limiting global warming to 2.0 °C. The fact is that the global warming rise reached a plateau with the increasing levels of anthropogenic greenhouse gases rise in the atmosphere. The global temperature pattern presents hiatus for the last nearly 19 years. By keeping the Paris meet people started bringing out papers after paper saying that global warming has not stopped -- All these groups are pro-warmists. Among these, the three important theories are as follows:

- Firstly, instead of looking this as a part of natural cycle [which we have presented long back with 60-year cycle similar to Indian Southwest Monsoon rainfall cycle, people started telling the temperature is hidden in the Ocean. Several papers were published on this theory in reputed Journals after peer-review;
- Secondly, after the Ocean theory, countering this argument later NOAA manipulated the data to erase the hiatus;
- Thirdly now, K. E. Trenberth says "Internal climate variability masks climate warming trends".

Now they accepted the existence of natural variability in climate, principal component of Climate change. They are cyclic in nature and they are beyond human control and we have to adapt to them. This is what our forefathers did in terms of agricultural practices and preserving water resources. Extremes in precipitation and temperature are part of this cyclic pattern only. This is superposed on the trend caused by human actions. At local/regional levels the other important component of climate change that plays the vital role is ecological changes associated with changes in land use & land cover such as urban heat-island effect, cold-island effect, deforestation & mining effect. They play role in the modification of climate at local/regional levels but they play no role at global level and yet global temperature curve includes this data. However, ground based observational network covers only around 20 to 25% of the globe. Because of this the temperatures derived through balloons & satellites present far over than the ground based data. This shows a rise of 0.25 °C since 1951 to date and off this around 50% is global warming. That means it is insignificant to influence nature, such as sea level rise. However, this is controlled by natural temperature variability of Ocean like PDO, Southern Oscillations, ADO, etc. These are modified by localized/regional human actions along the coast – building infrastructures along the prohibited zones, destruction green belt – protective walls of tidal fury --, drilling for water, oil & gas, sand mining, silting at the mouths of rivers entering the sea/oceans, etc.

The author's views were presented in Allvoices titled "Look climate change issues in Indian perspective rather blindly following IPCC". This is given as follows:

"Renowned environmental activist and Convener, Forum for a Sustainable Environment Dr. S Jeevananda Reddy suggested that there is a need to look in to climate change issues in Indian perspective rather than blindly following Intergovernmental Panel on Climate Change (IPCC) model based hypothetical studies with low scientific integrity, which have their vested interests. In a letter to the Indian minister for environment Mrs. Jayanthi Natarajan, he cautioned that if we go on concentrating our research on IPCC type research in our Indian institutes we achieve nothing except waste of public money. In this regard he discussed on two articles published with regard to IPCC release its brief on AR-5 report - 2013. He referred to Krishna Kumar Kanikicharla, climate scientist at Pune's IITM and one of the drafting authors of the IPCC report, told TOI there was "growing evidence" of the impact of the climate change on monsoons in South Asia and the tropical cyclone system in the Bay of Bengal. Kanikicharla, said: "There is strong hint that the duration of the rainy season would increase due to early onset of monsoon. The quantum of rainfall will also increase during the later part of this century." He said though the Indian summer monsoon circulation will weaken, rainfall will increase due to higher atmospheric moisture resulting from a rise in temperatures. In its projection for South Asia, the technical summary of the report clearly points at "enhanced summer monsoon precipitation and increased rainfall extremes of landfall cyclones on the coasts of the Bay of Bengal and Arabian Sea".

Prashant Goswami, chief scientist at Bangalore's CSIR Centre for Mathematical Modeling and Computer Simulation and one of the lead authors of the IPCC report, admitted that these conclusions were based on climatic projections that were not as firm as those made at a global level. "These uncertainties increase as you go to smaller scale (from global to regional or from regional to sub-regional levels). But one has to have an element of faith in these conclusions which are based on well-defined scientific methodology," Goswami told to TOI. As far as global temperature is concerned, the summary report said it was likely to rise by 0.3 to 4.8 degrees Celsius by the end of the century. Sea levels, it said, could rise by 26-82 centimeters by the end of the century.

As Unnikrishnan, chief scientist of the National Institute of Oceanography, Goa, said policymakers would have to rely on the global sea level rise projections to draw up their strategies to deal with the emerging scenarios. [This is Report 2 given below].

Dr Jeevananda Reddy said that his observations run like a street vendor's pep talk. It says the temperature may go up by 0.3 to 4.8 degrees Celsius and sea levels may rise by 26-82 centimeters asking us to have faith in those conclusions. Science does not work on faith but survive on facts and figures. You can see the range of variations in temperature and sea level rise. It is un-natural way of telling as far as science is concerned. This has no significant meaning. Also, they clearly accepted this by saying that "these conclusions were based on climatic projections that were not as firm as those made at a global level". This has no scientific sanctity.

As for as Indian Monsoon is concerned, he said, they follow a systematic variation and thus, with this temperature and agriculture follows [see my book published in 1993 "Agroclimate/ Agrometeorological Techniques: As applicable to dry-land agriculture in developing countries" and later written for India on the request of a publisher "Dry-land Agriculture in India: An Agro-climatological and Agro-meteorological perspective" in 2002 and "Green" Green Revolution: Agriculture perspective of Climate change published in 2011.]. Let me give few points here:

- From my paper published in IMD Journal in 1975 – the dates of onset over Kerala Coast present a 52 year cycle – see figure given below;
- All-India Southwest Monsoon data from 1871 to 1994 [published by IITM] showed a 60-year cycle – see given below figures. The same is evident in USA Hurricanes data – Typhoons of Pacific Ocean present opposite pattern. Anybody can check this by simply plotting ten years averages;
- Cyclones in Bay of Bengal follow 56 years cycle similar to Andhra Pradesh Northeast Monsoon rainfall [Southwest and

Northeast Monsoons rainfall follow opposite pattern but annual rainfall presents a 132 year cycle];

- North India's temperature in summer and winter follow the Western disturbances pattern – I published an article in IMD journal in 1978;
- Finally, all the conclusions drawn by IPCC are of hypothetical in nature.

He also said that marine scientists warned on Thursday that our oceans are declining more rapidly than previously thought — becoming ever warmer and more acidic and losing oxygen at an alarming rate. "It is the simultaneous occurrence of the 'deadly trio' of acidification, warming and de-oxygenation that is seriously affecting how productive and efficient the ocean is". It also warned of a fast drop in dissolved oxygen in seawater — with predictions of a 1-7 per cent decline by 2100 caused by the effects of global warming and nutrient runoff into the sea from agricultural fertilizers and sewage. The new report said the average upper ocean temperature has increased by 0.6 degrees Celsius (1.08 deg Fahrenheit) over the last 100 years. Further warming will likely cause the disappearance of Arctic summer sea ice, further oxygen depletion and the release of Earth-warming methane trapped in the Arctic seabed. He said that from this it is clear that the oceans are declining rapidly due to three reasons – global warming, pollution caused by runoff from agriculture fields, and Arctic sea ice loss.

Dr Jeevananda Reddy said that global warming in the last 17 years is lull. The sea temperature and carbon dioxide present a systematic variation [see figure given below] with the time like Sunspot cycle see my book "Climate change: Myths & Realities" (I-2008 & II-2010). So, warming and cooling of ocean water is a natural phenomenon operating over centuries. If we look at the data of seasonal variation of sea ice year by year at Arctic [North Pole] and Antarctic [South Pole] the year to year variations of seasonal trends are above the average at Antarctica and below the average at Arctic in the last decade. They fall within the long term standard deviation limit except 2012 Arctic seasonal pattern that falls below standard deviation shade. In statistical terms even up to twice the standard deviation the extremes occur. The main issue that needs more study is the pollution impact on oceans. With the green revolution technology of chemical inputs, runoff from agriculture farms entering oceans and creating dead zones – zero oxygen – like thousands of square kilometers in Gulf of Mexico. Also there are several other sources of pollution contaminating ocean waters.

Before IPCC AR-5 report briefing/press release another report executive summary on “Climate Change Reconsidered II: Physical Science (CCR-II) was released by NIPCC – Non-governmental International Panel on Climate Change — at a news meet in Chicago on 17th September 2013. He said that before release they asked his comment on the report and after release they asked his comments on three queries which they consolidated from observations on their online post in a Canadian Newspaper. He responded on the two. They are as follows:

1. “According to the Commonwealth Scientific and Industrial Research Organization, sea levels on average have been 20 centimeters higher than they were in the 1880s. Water levels are rising at 3.2 millimeters per year, which is double the 20th century trend.” -- His response: In the past on several forums I made the following observations but nobody come forward to respond my question: (1) San Francisco Airport is in the Ocean waters. Was there any rise in sea water level at this point? — so far no change [I visited several times]; (2) In Italy along the coast, traditional and historical wells showed no change in sea level; (3) It is generally argued that the sea level raise submerged the Sundarbans at the mouth of Ganga River in the Bay of Bengal and villages along Bramhaputra River in Bangladesh — all this is inaccurate statements. Here the silt deposition from these two mighty rivers causing the submergence of Mangroves and hutments; (4) Also, destruction of Coastline for commercial activities and destruction of Mangroves allowed sea water entering into nearby villages and thus people are attributing it to sea level rise; (5) In several parts, sea sand is lifted for other uses and thus causing the sea water entering in to nearby areas; and (6) Boston area was filled [artificially built] and with the storm surges it looks something happened. Also, sea temperatures are higher in some places and some other places they are cool associated with general circulation patterns existing generally in those zones. These infect the sea level temporarily — natural variations. Unless we analyze region-wise by taking into account on ground

variations in the light of general circulation, natural variations & physical impacts, all model estimates are speculative in nature and create sensation that pickup media.

2. "Ocean acidification [due to higher $CO_2$ levels] and warming are causing species like jellyfish to overrun other species in the sea and bleaching and decreasing growth of coral reefs"

   His response: In the oceans, temperature change and along with carbon dioxide change is a historical fact. They clearly indicate a clear cut cyclic variations — I put this in my book "Climate Change: Myths & Realities" published in 2008. As this is not new the life in the ocean automatically adopt to these variations. There is a need to look into other man induced pollution factors in to ocean waters angle to understand their impact on coral reefs as well other species in the ocean. Our research, therefore, must direct to understand the cause and effect mechanism. With the preconceived ideas we go nowhere. This is exactly what is happening now. This is bad science — though many a times such studies are published in reputed journals, as editorial boards work "you scratch my back and I scratch your back".

3. "The physical principles behind climate change are simple, well known since the 1800s, and won't wait for this fake debate to end". His response: This point I did not understand, here climate change means a de facto global warming or all issues pertaining to changes in climate. As per global warming there are no clear cut physical principles. We can see in the literature every other day a new issue comes up. Now, even IPCC also changed and telling urban-heat-island affect also contributing to global warming by about 10% — initially they did not accounted this –. There are several other issues like natural variation component, solar components, ecological changes components [that influence local and regional climate], physical impacts on ecological sensitive zones like ice, etc. It is not clear on the link between carbon dioxide and temperature. It is a statistical link. It varies with the accumulation of data series period and it does not explain the physical relationship.

The main component that at present influencing climate and extremes is the natural variations systematic & irregular variations and ecological changes such as changes in land use and land cover effect local and regional climate and thus change the natural variations to that extent. At Antarctica the sea ice melt in the last six years is above the mean while it is below the mean in Arctic area.

The following are few other issues presented at international forums:

Why don't they just show the annual temperature graph? They always thought it most appropriate in the past. Why do they begin this graph in 1850? Anthropogenic Climate Change, according to their report, could not have started before 1950 (a hundred years later). So why aren't the chart, and all the ballyhooed temperature increase, started in 1950? If AGW didn't start (couldn't start, actually, as we hadn't yet added the $CO_2$ to the atmosphere), what caused the temperature rise starting in 1890? Whatever that cause is, what made that cause stop and suddenly change to $CO_2$ - caused in 1950? Models present a monotonic increase as the models are built with some preconceived notions that means what you are able to think only goes in to the model. Nature in fact acts differently based on several unknown and known factors. So, the real pattern goes on changing with the time. However, this is affected by scientists' manipulation of data and changes in land use and land cover as weather is highly sensitive to such variations. Also, we are experiencing changes due to man's physical actions and manipulation of terrain. These changes are alarmingly increasing with population growth and changing technologies to meet their lifestyles and needs. In global models such factors taken in to account rarely. Also natural disasters do change the terrain conditions. And above all, the natural variations are not the same over different parts of the globe. Instead of talking on a generalized term "global warming", it is more appropriate to use the term "regional climate changes". This is more useful in regional planning. Global temperature data is manipulated, global carbon dioxide data is manipulated. Meteorological data collection is changing with the time over land and in/on oceans with network and instrumentation changes. Satellite data is highly manipulated!!!

In 70s & 80s, before encroachment of global warming in to climate studies, before starting analyzing rainfall data by clubbing the rainfall of different rain gauge stations, we used to homogenize the rain gauge stations in terms of rainfall patterns. This helped to understand the climate of different parts in a region. This helped better interpretation of the results so analyzed in terms of agriculture or water management. It appears this is lacking in IPCC studies and counter positive and negative trends giving false notions or predictions.

Other important issue is IPCC talks of probability. Probabilities relevance with data but not number of people accepted it. This is bad science. In science the phrase 70% probable or 90% probable had definite meanings. They imply controlled trials, they imply numerical quantitative information objectively assessed. If you ask the IPCC they will tell you that when they use the term 95% probable it is based on the expert opinion of a group of people gathered around a table. It is completely wrong to use probability terminology to describe what is albeit an expert opinion.

The global yearly temperature pattern presents a 60 year cycle sine curve with 30 years above the average and 30 years below the average varying between –0.3 to +0.3 degrees Celsius superposed on a linear fit – according to this by 2100 the rise is around one degree Celsius – see figure given below. However, year to year variations, known as irregular variations are common. Anybody can try this. Nothing is secret here. Look at WMO (1966) Manual "Climate Change". You can take the published temperature data.

The linear part consists of several components that are influencing globally. Here you get good probability unlike IPCCs probability based on number of persons agreed. We must not forget the fact that when we talk of precipitation, we must first demarcate homogeneous zones in region based on individual rain gauge stations data. In a homogeneous zone the rainfall presents similar patterns. If we mix non-homogeneous zones and take the average, the average patterns do not reflect the realistic pattern and mislead the interpretation and the consequent impacts. Take for example: Durban in South Africa, Mahalapye in Botswana & Catuane in Mozambique all three present W followed by M pattern of 66, 60 & 54 year cycles. When we add these three get meaningless result for agriculture purposes – same will be the case with global average temperature. The author of the book did homogenization studies for few countries including Brazil."

### 4.3.3 Water Resources

Water resources are basically defined by the process of hydrological cycle [**Figure 1.3**]. Water comes on to the Earth through precipitation and melting of ice/snowfall. They create river flows as well groundwater flows. Some of this is stored on surface reservoirs and some goes into sea. Two-thirds of the Earth surface is covered by water. Some part of the water in turn reaches the atmosphere through evaporation from the water bodies and soil; and evapotranspiration from plants and greenery. It forms an important greenhouse gas that regulates the greenhouse effect in the surface-troposphere and thus temperature.

Evaporation and evapotranspiration are defined by several factors that includes temperature regime of the region. So also is the case with melting of ice/snow. With the increased temperatures accelerates glaciers retreat and thus increases snow melt. This helps the sea level rise; and over time, rivers that are being fed by these glaciers will dry-up. "This is the theory" but in nature they are controlled by systematic variations in climate. This process affects agriculture and ecosystems. However, pollution and deforestation processes impact water resources severely with the progression of the time in several ways. The systematic variations play vital role on the river flows, groundwater recharging and soil moisture. **Table 4.3** presents the frequency of occurrence of floods in the three northwest Indian rivers during the below and above the average periods of 60-year cycle of All-India Southwest Monsoon precipitation data series. The flood data was taken from "State of Environment Report, India – 2009 of MoEF". This report states that the frequency of floods in India is largely due to deforestation in the catchment areas, destruction of surface vegetation, changes in land use, increasing urbanization and other developmental activities. This is not correct as the high and low frequency of flood occurrences in rivers are associated with above and below the average precipitation cycle parts. In each cycle average precipitation decreases from the above the average 30 year part to the below the average 30 year part. The minister of Earth Sciences used the data of second cycle and informed the parliament that precipitation is decreasing. If you use last part of the second cycle and first part of third cycle, it looks as if precipitation is increasing [**Figure 3.16**]. That means truncated data series misguide users if the data series present a cyclic pattern; and thus lead to misleading inferences on cause and effect. This is exactly what happened with the two cases mentioned above. Also, what happened with the Brijesh Kumar Tribunal on Krishna River water distribution among the three riparian states [**Reddy, 2016**], while doing so the tribunal identified a period that represents the above the average precipitation period in which flood frequency is expected to be more. Prior to this, Bachawath tribunal took the data covering both above and below the average periods. These two periods thus showed a difference of 185 mm on an average level. That means Brijesh Kumar tribunal selection of better rainfall period affected negatively the downstream Andhra Pradesh and positively upstream Karnataka and Maharashtra. Therefore, to mitigate the impacts of droughts and floods, governments must look for ways and means of: storing flood waters; interlinking of rivers; development of groundwater recharging mechanism – at present we are extracting more than the recharge; implementing better land and water management practices – it is important in dry-land agriculture; implementing micro-irrigation techniques with groundwater; stop using high water intense crops with groundwater irrigation – with reduction in groundwater power consumption increases with the time, etc.

**Table 4.3** Periods of three cycles and frequency of occurrence of floods in rivers

| Periods | Average precipitation [a – b], mm |
|---|---|
| – Cycle 1 [1867/68 - 1926/27] – 873 - 828 | |
| – Cycle 2 [1927/28 - 1986/87] – 868 - 837 | Minister of Earth Sciences used |
| – Cycle 3 [1987/88 - 2046/47] – 860 - | ??? |

| River | Period | Frequency | Climatic cycle |
|---|---|---|---|
| Chenab | 1962-1990 | 1 in 9 years | (a) Below the average cycle |
| | 1990-1998 | 1 in 3 years | (b) Above the average cycle |
| Ravi | 1963-1990 | 1 in 14 years | (a) |
| | 1990-1998 | 1 in 3 years | (b) |
| Beas | 1941-1990 | 1 in 8 years | (a) |
| | 1990-1995 | 1 in 2 years | (b) |

Let me explain ambiguous results of truncated data under cyclic variation data series: **Figures 4.4a & b** present the trends with full data set and with truncated data sets with rapid increase ['a' temperature anomaly data and 'b' with the North Atlantic SST]. This type of trends show accelerating in the later part, a fictitious conclusion as this accelerated increase is part of one arm of the cyclic sine curve. If the data is random one the conclusion may be correct and not with cyclic variation data as that seen in the figure. That is truncated data series of systematic variation give spurious inferences – it may present increasing trend or decreasing trend.

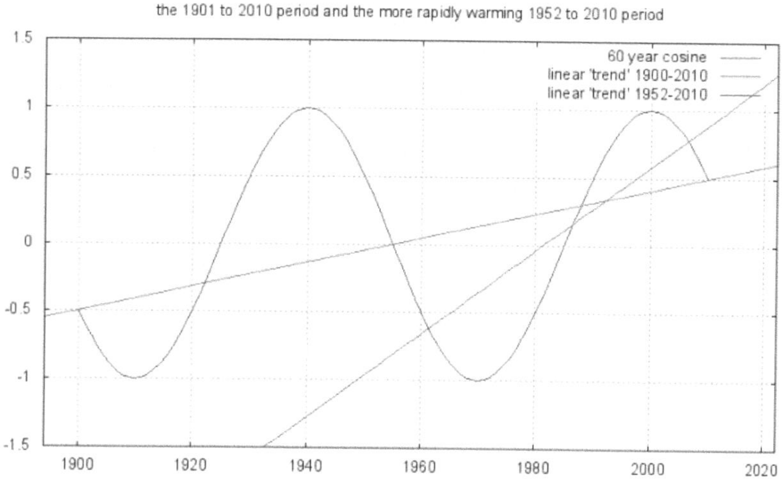


[a]

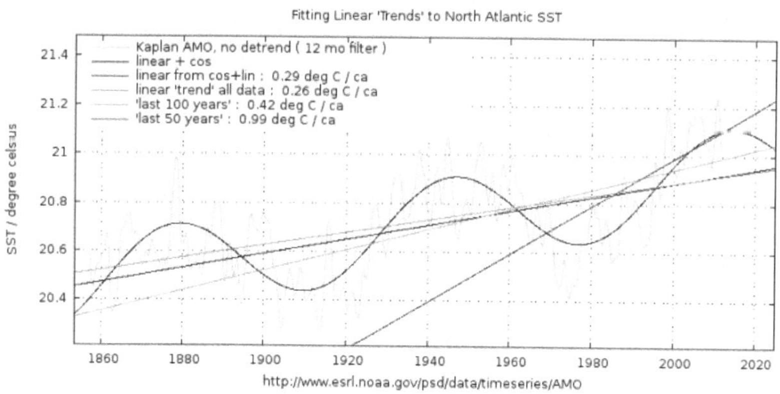


[b]

**Figure 4.4** Fitting linear trend to [a] temperature anomaly data & [b] North Atlantic SST

The longer data series represent symmetric system of above and below the average and the shorter time series presents an asymmetric system of above and below the average with below the average period wherein data points are dominating/more. Naturally the trend showed higher angle or more accelerating [provides false alarm] than the former. We generally call such fittings as biased.

**Fresh water resources:** Key risks at the global scale, IPCC observed "So far there are no widespread observations of changes in flood magnitude and frequency due to anthropogenic climate changes [means by global warming] but projects imply variations in the frequency of floods (medium agreement, limited evidence). Flood hazards are projected to increase in parts of south, southeast and northeast Asia, tropical Africa, and South America (medium agreement, limited evidence). Since the mid-20$^{th}$ Century, socio-economic losses from flooding have increased mainly due to greater exposure and vulnerability (high confidence). Global flood risks will increase in the future partly due to climate change (medium agreement, limited evidence). There is little or no observational evidence yet that soil erosion and sediment loads have been altered significantly due to changing climate (medium agreement, limited evidence). However, increases in heavy rainfall and temperatures are projected to change soil erosion and sediment yields, although the extent of these changes is highly uncertain and depends on rainfall seasonality, and land cover, and soil management practices. **Chapter 3** of the book dealt the issue of natural variations in the precipitation.

**Food security and production systems:** Though crop production relates to the fresh water resources, IPCC observed "The effects of climate change on crop production are evident in several regions of the world (high confidence). All aspects of food security are potentially affected by climate change, including food access, utilization, and price stability (high confidence).

**Asia:** IPCC observed "Warming trends and increasing temperature extremes have been observed across most of the Asian regions over the past century (high confidence). Increasing numbers of warm days and decreasing numbers of cold days have been observed, with the warming trend continuing into the new millennium. Precipitation trends including extremes are characterized by strong variability, with both increasing and decreasing trends observed in different parts and seasons of Asia.

This is exactly what I was advocating (high confidence) since long wherein the observed changes in nature are mainly associated with physical impacts due to human action and natural variations; and not with the anthropogenic global warming. Also, so far the increase of anthropogenic global warming is insignificant to influence nature and weather systems. We are working in a climate system that showed high seasonal and annual variations. We can see this from Climate Normal books – this aspect was discussed in **Chapter 1**. We need to make better and efficient use of the available resources. We must choose technologies that will not destroy our natural resources. We must look at pollution and not at anthropogenic greenhouse gases.

Unfortunately, IPCC and other agencies are attributing all ills of the society to global warming, a phenomenon that is insignificant to influence nature. The issues that are raised under water resources and agriculture by IPCC are nothing to do with global warming but they are primarily related to poor governance and human physical impacts on nature. We must adapt to natural variations in climate. This is what our forefathers did. We need to control pollution [air, water, land & food] and reduce the wastage in food and water resources through better management. We must use local technologies that help in this direction.

### 4.3.4 Ecosystems

Ecosystem is different from ecological changes. An ecosystem is a collection of communities of organisms and the environment in which they live. Ecosystems, like in traditional agriculture, are adapted to a location and region. They survived centuries under vagaries of weather. That is, ecosystems are adapted to vagaries of weather *in situ* over several centuries. Glaciers retreat and sea level rises affect the ecosystems. However, ecosystems are mainly affected by factors other than climate change such as deforestation, alien species invasion, insects-pests-diseases entry, pollution, natural disasters, etc. After 1960's the new agriculture technology, more particularly genetically modified seed technology also contributing to major destruction of on land biodiversity.

Modern agriculture GM technology; Physical impacts tourism, pilgrimage, sports; Infrastructure & Pollution major causes;

Population growth and technology based lifestyle; Species have certain degree of tolerance to normal climate of the region survived for centuries under vagaries of weather; Deforestation, alien species invasion, insects-pests-diseases entry, pollution Acid Rain, natural disasters, etc.

The increased temperature impacts indirectly “glaciers retreat and sea level rise”. IPCC and Al Gore projections of sea level rise and glacier retreats were based on the model simulations by assuming the temperature rise of 3 to 4 °C by 2100. So far the observed increase over the average is less than 0.15 °C [global warming component] and based on the observed data series the projected expected temperature reach 0.20 °C by 2100 only. Thus, the impacts of global warming inferred by IPCC and Al Gore are several times higher than that from the observed data expectations.

IPCC report says due to their predicted global warming trend: Habitats of many species will move pole-ward or upward from their current location; globally by the year 2080, about 20% of coastal wetlands could be lost due to sea level rise; glaciers and ice caps are projected to continue their widespread retreat during the 21st century; etc”. In the case of Himalayan glaciers, IPCC predicted that most of them disappear by 2030. However, IPCC has withdrawn this conclusion saying that this is not based on authentic data source. Very recently ISRO, Ahmadabad issued a report based on the satellite observations, wherein they stated that some glaciers are retreating, at some no change and some others are growing. Similar observations were made with sea level: in some zones sea level has gone up, at some zones it has gone down and at some other zones no change. That means they are not associated with global warming phenomena as such or for the matter of fact, there is no global warming as such except that this is a sum and total of local-region specific changes.

Al Gore claimed that the latest research showed that the Arctic could be completely ice-free in five years. Al Gore now accepted this as a mistake and withdrew his conclusion “Arctic could be completely ice-free in five years”. Arctic sea ice melts and refreezes in an annual cycle, reaching its lowest point in late summer, and its highest point in late winter. After the satellite measurements in 1979, in summer the lowest was observed in 2007 and the highest was observed in 1996. However, a decreasing trend was reported since 1979 a fall of 10% per decade. Over this there are high year to year variations in association with the variable weather patterns. However, the winter pattern is more or less the same. Britain's Times Comprehensive Atlas of the World "The Greatest Book on the Earth", presents that Greenland lost permanently ice cover by 15%. The scientists want publisher to refreeze Greenland as this is around 1% only. The agency agreed to rectify the error. All these clearly reflect that this is not associated with the global warming. In Greenland zone with volcanic activity new islands are farmed and older ones are disappearing.

Reports say that "the beetle is now wiping out a whole ecosystem high elevation white bark pine forests that saw only limited beetle out breaks during warm spells in the past". Warmer climates are created by deforestation, forest fires and natural cycles in temperature, weather aberrations and heat advection, etc. However, scientists have not defined the share of long term temperature pattern, ice ages part on such impacts. This is like story of Indian apple.

### 4.3.5 Unusual Events

In the past one decade, there were heated discussions on extreme weather events world over. Unfortunately, "media" only picks up the material from such discussions to sensationalize the issue so that the "media" gets hype. The basic question we can ask is "When was the period in human history in which some person, in some location or other, wasn't feeling that some recent day, month or season has been very unusual in some dry, wet, hot, or cold way?" That is, extreme weather events are part of natural irregular part of systematic part of climate change. Recently an article "Pacific hot spells shifting as predicted in human-heated world" was discussed in New York Times. One of the authors McPhaden of the article replied saying that "If all we have was data, the interpretation would be ambiguous as to what causes what. We can't be 100% certain that this interpretation is correct because the data are limited and models are imperfect".

Presenting a new NOAA report on 2011 extreme weather, Dr. Karl said that extremes of precipitation have increased as the planet warms and more water evaporates from the oceans. He also said models suggest that as carbon dioxide builds up in the atmosphere and heats the planet, droughts will increase in frequency and intensity. "But it is difficult and unlikely to discern a human fingerprint, if there is one, on the drought record of the United States," he said. From these it is clear that the scientists try to attribute extreme events to global warming from air and to cover up this finally they say such events also were present in the past. But they were careful not to blame humans for this year's rash of deadly events, saying that in some ways weather patterns were returning to those seen at the beginning of the last century. "Looking at long-term patterns since 1980, indeed, extreme climatologically and meteorological events have increased," said Thomas R. Karl, director of NOAA's National Climatic Data Center. "But in the early part of the 20th century, there was also a tendency for more extreme events followed by a quiet couple of decades." He did not do much on the why that scenario. He did not look at systematic variations in met parameters. They left to the readers.

Compilations by the Vermont Historical Society show the astounding devastation in that state on November 3 and 4, 1927, after the remains of tropical storm produced a 36-hour deluge. Review the research from the Northeast shows that far more intense scouring floods than this event have occurred throughout the period since the end of the last ice age. Take the history of Nile River in Egypt. Trend analyses for the 572 eastern United States gauging stations provide little evidence at this point (2009) for increasing floods peak distributions associated with human induced climate change. In India river floods are associated with rainfall cycles. Systematic precipitation patterns exist all over the world. These are modified based on local ecological changes, like deforestation. In India the heat waves in summer and cold waves in winter are common phenomena that are associated with the western disturbances, a system of general circulation pattern in a given year.

Unfortunately most of the political-scientists sensationalize the issue with their copy-cat rhetoric's from IPCC – Al Gore. The Indian scientists are also travelling in the same boggy!! Therefore, there is an urgent need to look in to such scenarios by national organizations using historical data along with the ecological changes that took place over different parts/regions of the globe with the passing of time. Unfortunately, some scientists just because of their position become "top scientists" make sensational press releases from the air. Such scientists must stop such pronouncements. Only confirmed, based on the history must go to press as otherwise the research lead nowhere – Russian abnormal heat was attributed to global warming but later withdrawn saying it is natural variation. Let the nations create a data bank of unusual events that will help to clear this type of misgivings that affect national planning and economy. All such issues come under irregular variations or year to year weather aberrations. These are different from climate change. In fact when the author of the book served the IMD, he used to prepare a report on unusual events of that year. This was later released by IMD headquarters every year.

Nicholas St. Fleur contributed the following report to the New York Times [5th November 2015]. He raised some basic questions like: "Did climate change cause that heat wave, that hurricane, that drought" Then he presented a report that "based on a collection of studies examined extreme weather events last year, including drought, floods and storms, to look for signs that climate change was a cause or contributor. The papers are part of a broader effort to recognize the effects of climate change as the world warms, and to tease out those factors from other possible causes of extreme events. – Here he used climate change as defacto global warming.

Climate change is often discussed in terms of predictions about what may happen in the next 100 years or more as average global temperatures rise. But an emerging field of science is dedicated to discerning whether climate change is already having effects, and what they might be. The set of 32 studies published Thursday examined 28 extreme weather events in 2014; it is the fourth in a series of annual reports, and appears in The Bulletin of the American Meteorological Society."

He further quoted that "The studies this year are pretty evenly split, about 50-50, for those that did and did not find a role for climate change in the event's likelihood or intensity," said Stephanie Herring, lead editor of the report and a National Oceanic and Atmospheric Administration climate scientist, in a news conference. "The strength of the climate change signals that we see really varies based on the type of event being looked at."

The papers suggest that "human-caused climate change greatly increased the likelihood and intensity of heat waves" in some regions, including Argentina, Europe, China, the Korean

Peninsula, Australia, and the northern Atlantic and Pacific Oceans. The researchers also say that climate change probably had a role in the unusually large number of tropical cyclones that hit Hawaii last year.

The scientists, however, did not discern the influence of climate change in every event, nor did they see it playing a consistent role in certain types of events, like drought. According to the papers, which run to nearly 200 pages altogether, some droughts appeared to have more to do with population growth and government policies than rising levels of carbon dioxide and its effects. The researchers, however, did conclude that the Syrian drought was made worse by a lack of rainfall linked to climate change. Is it global warming? Then his conclusion is erroneous.

"It is by no means a prevailing one-story-fits-all-events type of approach to this," said Martin Hoerling, a co-editor of the report and a NOAA meteorologist, at the news conference. "It does require a specific analysis of each case on its own merit." No strong link was established between climate change and the wildfires last year in the American West, or to the unusually tough Midwestern winter, which was tied more strongly to a pattern of unusual winds in the tropical Pacific. The authors added, however, that even where no link to climate change was detected, "The failure to find a human fingerprint could be due to insufficient data or poor models." The authors emphasized that they were not describing direct cause and effect, but an analysis of probabilities based on computer models and other factors. "The analysis contains uncertainty" said Hiroyuki Murakami, an author of the paper on Hawaiian hurricanes and an associate research scholar at Princeton who works at the Geophysical Fluid Dynamics Laboratory at NOAA. The analysis by Dr. Murakami and colleagues took into account many factors, including natural variability and phenomena like El Niño that play roles in hurricane formation. But the "signal" of human-caused climate change was clearly present in the models, he said, so "we still see that global warming increased the probability of tropical cyclones in Hawaii." This statement is inaccurate. In USA annual temperature anomaly data series [raw] from 1920 onwards showed a 60 year cyclic pattern.

The efforts to evaluate the effects of climate change are important for planning responses to extreme weather and a changing world, said Heidi Cullen, an author of one of the studies and chief scientist of Climate Central, a news and research organization in Princeton, N.J. "Communities, businesses and governments affected by extreme events are increasingly asking for objective assessments of the causes," she said. Dr. Cullen also leads the World Weather Attribution programme, an international coalition formed to quickly determine climate links to extreme weather events. "In some instances, scientists are now able to quantify the relationship between an extreme event and long-term warming trends, rather than consigning the public to reliance on dueling speculations not grounded in science," she said. The meteorological society reports do not sit well with some climate scientists who remain wary of the focus on tying climate change to specific events rather than taking a broader view. Michael E. Mann, a climate expert at Pennsylvania State University, said the current level of warming of about 1.5 degrees Fahrenheit around the globe (and somewhat more over North America and the Arctic) "has fundamentally influenced all meteorological events," not just those that get written up in a study. Dr. Mann said the work and the debate about it were useful. "If anything, this particular debate underscores that the question is no longer whether there is an influence of climate change on extreme weather events. The debate is simply over the magnitude and extent of that influence.

**2015 winter in Eastern USA & Montreal in Canada:** The warm weather on the eastern half of USA 2015 winter has prompted many climate alarmists blame it on global warming/climate change/climate disruption even though in 1955 a similarly warm and record breaking warm winter occurred well before $CO_2$ emission scenario.

On the 24th December 2015 Montreal's Christmas-eve record breaking temperature matches Los Angeles. Montreal saw a record snowfall for December 29, 2015 with 39.2 cm of snow blanketed the city. Environment Canada confirmed the record, 30.5 cm of snowfall in December 29, 1954.

**Hail storms:** During the pre-monsoon season [April-May] in India, hail storms are common in India – the clouds are dark and thick. In every year one part or the other they occur. On 30th April 1888 a severe hail storm devastated the farming town of Moradabad in UP, killing 230 people and many more animals. Also 16 people were killed in nearby Bareilly. The size of the hail was recorded to the size of orange and on ground accumulated to around 2 feet height. Generally hail is accompanied by strong winds that affect the thatched houses, common in rural India.

**Fallacy:** The main fallacy in declaring an even as unusual is based on the past records. The meteorological records started only around 1850s and that too at fewer locations. With the progression of time they increased in number covering wider space. Prior to 1850 unusual events only some documents narrated but they are rarely available to public to consult like met data. With the modern political set up the lapses committed by the government bodies put on to unusual event and the media give them hype.

In all the unusual events they used invariably the word "climate change" but while discussing the event they talk of impact of temperature, which refers indirectly to global warming. As I presented earlier, global warming is different from climate change. The unusual events are irregular variations part of systematic variations in climate change that are primarily associated with the general circulation patterns of a region – part of climate system. The global temperature rise is not global. Major part is local and regional. Even here, the rise is part of data manipulation and is not seen in satellite & balloon data series. Even the satellite data is manipulated by withdrawing original data from internet.

Before sensationalizing an event, the authors must do some basic study and also the media should stop picking up some sentences to give hype to the readers. Local governments must look in to such sensationalizing reports with ground realities and clarify if any to media. It has become a fashion to use the word "climate change" as an adjective to get hype to the work.

Even, WMO Secretary General in his World Meteorological Day statement released in 2014 observed that the drought conditions were associated with global warming in the Southern Hemisphere in 2013. On this the author of the book sent his response saying that this was predicted [drought is expected in 2013 according to extrapolated predicted curve] in my studies on the natural cyclic variation in Southern Hemisphere rainfall series [**Reddy, 1993**]. Also, heat waves and cold waves particularly related to general circulation patterns over different parts and they are part of intra-seasonal and intra-annual irregular variations [**Figure 1.2**; **Reddy & Rao, 1978**]. The main ambiguity in explaining the cause and effect syndrome, we must clear the fact that global warming is not climate change. If we correct this, we get correct cause and effect relations.

## 4.4 Impact of Human Greed

In a natural weather conditions scenario, many a time human greed plays vital role in the loss of human lives and destruction of property and natural resources.

**Flood disasters along the Riverbeds**: Droughts and floods are common to India for that matter world over. Each year one part or the other in India experiences the floods and droughts. The severity of destruction changes with the time of the year, the terrain, with the population growth, and growth in infrastructure. However, with the violation of existing local, state and national laws the destruction is aggravated. The author of the book wrote as well presented in several TV discussions in Hyderabad/India on the causes of flood disasters in Uttarkhand in June 2013 [**Figure 4.5a**] and Jammu & Kashmir in 2014 [**Figure 4.5b**]. **Figure 4.5c** presents general scenario in 2015. The author of the book countered the arguments, attributing the causes to climate change a de-facto global warming and explained the reasons for such scale of disasters as "manmade", and nothing to do with global warming, due to apathy of government agencies, who were unable to control the illegal construction activities along the river beds. This has lead more than 10,000 deaths in Uttarkhand floods along the property losses running in to thousands of crores. Same was the case even with Jammu & Kashmir floods. The recent flood disasters happened mainly because residences and commercial establishments had been indiscriminately raised by unsuspecting people in active floodplains. Much of this could have been avoided had the local governments/municipal authorities would have implemented local guidelines on construction activities along the riverbeds. The official collected their hefty bribes and as a result the illegal construction activities flourished along the riverbeds. Same is the case with November-December 2015 Tamil Nadu floods. In the capital city Chennai the rainwater channels, rivers, water bodies were converted in to concrete jungle. This has resulted severe floods, though similar rainfall was experienced earlier with less or no flooding. This is also the same in Nellore in Andhra Pradesh. Though, few years back I submitted a note in writing [oral presentation] to disaster Management group on the problems for causing floods in urban areas. But unfortunately, the committee hired an IIT professor who was not interested in such thing but only interested to attribute the flooding to global warming.

**[a]** Uttarkhand floods of 2013

**Note**: Uttarakhand floods had claimed thousands of lives in 2013. That floodplains need to be protected and a strict regimen of dos and don'ts need to be put in place

**[b]** J & K floods of 2014

**Note:** The 2014 floods in Jammu & Kashmir claimed more than 200 lives and caused losses in excess of Rs 1 lakh crore

**[c] 2015** floods

**Note:** In the last one decade, India has faced flood disasters in Maharashtra, Bihar, Uttarakhand, Assam, Jammu & Kashmir and Odisha that caused massive losses of life, property and infrastructure

[d] Chennai floods in 2015

People travel on a boat as they move to safer places through a flooded road in Chennai, India, December 2, 2015. REUTERS/Stringer

**Figure 4.5** Human Greed Impact on nature in India – [a] Uttarkhand Floods of 2013, [b] J & K Floods of 2014, [c] 2015 Floods, & [d] Chennai floods in 2015

Darpan Singh presented an article in mail online on 24$^{th}$ October 2015 titled "India's major cities face flooding disaster after riverbed protection plan is drowned in red tape for thirteen years".

He corroborated with my view points. The article observed that "Several Indian cities are in danger of being ravaged in floods as the Union Ministry of Environment and Forests (MoEF) has failed to implement a river floodplain policy a draft of which was prepared 13 years ago. In the last one decade, India has faced flood disasters in Maharashtra, Bihar, Uttarakhand, Assam, Jammu & Kashmir and Odessa that caused massive losses of life, property and infrastructure. Floods in Jammu & Kashmir in 2014 claimed more than 200 lives and caused losses in excess of Rs one lakh crore. A River Regulation zone (RRZ) draft policy which awaits government notification looks at dividing the area of the river floodplain into zones, the one closest to the river channel to be called "no development zone". Principal author of the of the draft policy Brij Gopal said: "When there is no space left for rivers to spread their waters, they will rise up and down your houses. First we have through structures like dams and barrages turned our rivers dry. Then we encroach into the dried-up spaces. When monsoon brings water, we call it 'devastating' floods. "The government liked a civil society idea of floodplain protection and asked a group of activists how this can be done."

**Chennai & Nellore floods of November/December 2015 [Figure 4.5d]**: Nellore to Chennai November 2015 floods are natural part and not due to global warming. The Times of India [29-11-2015] reports that "Prime Minister Narendra Modi on Sunday condole the deaths in Tamil Nadu due to incessant rain and floods, which has claimed at least 170 lives so far. Addressing the 14th edition of his 'Mann Ki Baat' radio programme, the Prime Minister drew the nation's attention towards climate change and global warming. "It was saddening to witness devastation and deaths during the festive season, but I have faith in the strength of Tamil Nadu," he said in his 27-minute talk, adding that the tragedy in the state was due to the impact of global warming. "Checking global warming is everyone's responsibility. Energy conservation is the best way to do it," Modi said. On this the author of the book sent the following mail to Hon'ble Prime Minister of India:

As a Prime Minister of India, Narendra Modi Garu should have verified the facts before attributing the floods to global warming. Unfortunately people are using the word very casually. In the Southern coastal zones receive heavy rains during Northeast Monsoon Season [October to December, post-monsoon season, temperatures in descending seasonal mode] that coincides with severe cyclonic activity. Let me present the highest rainfall in 24 hours [mm]/year and the highest per month [mm]/year for Nellore and Chennai from 1931-60 Normal [Red Book of IMD]:

Nellore – 444/1950, 357/1936 & 189/1902; 647/1920, 982/1915, 494/1946

Chennai – 234/1888, 236/1922 & 262/1901; 892/1943, 1088/1918, 699/1946

The November 2015 rains followed this pattern only.

The main problem here is that governments over the years allowed the people as well government agencies to destroy the rainwater channels and water bodies/rivers. This resulted the aggravating the impact for the same flood capacity. This is same in Uttarkhand in 2013, Jammu & Kashmir in 2014. Also, the August 2000 floods in Hyderabad.

To meet the greed, humans have destroyed the natural flow systems and now governments are putting the blame on global warming as it cannot defend against such onslaught by politicians and bureaucrats to protect themselves for wrong doings. Easy safe goat is for all ills [heat waves, cold waves, floods, cyclonic fury, etc] the easy prey is global warming. Climate change is used as defacto global warming, though natural variations are part of climate change for which we need to adapt to them. The Paris meet [COP21] is fight for getting their share from green fund of $100 billion per year.

The British Government officials have predictably blamed climate change to severe flooding which has afflicted England in recent weeks [December 2015]. But there have been also strong criticisms of river management policies. In Egypt, officials tried to blame flooding in Nile Delta in October 2015 on climate change. The Egyptian official was forced to resign. It seems unlikely any one in Britain or Europe will be forced to resign because of mismanagement of Britons waterways – drains. Even in India the scenario is the same. In all such human greed the easy escape goat is climate change.

On 6th December 2015 a huge storm "Goliath" poured 9-10" of rain [most of it occurring in three days] in a belt across the central United States, centered just Southwest of St. Louis in the Mississippi River Basin. 1982 flood was similar 2015 flood. Also, both floods were winter flood during an El Nino event. They should have been similar but it is not so. After 1982 there were large scale changes in the catchment area with construction activities and thus water level rise by several feet over 1982 flood level, resulting in the damage of 7000 buildings and several other damages and few people died. It is a manmade disaster.

## 4.5 Impact of Climate Change on Agriculture

### 4.5.1 Effects of Climate Change on Agriculture

The IPCC report observed that "Climate change will especially affect the livelihoods of people, agriculture the mainstay of the Indian economy, will see dramatic changes in yields, affecting people's right to food security. The Supreme Court of India on 1st February 2016 slammed the Gujarat State Government and some other state governments for not implementing the National Food Security Act and raised questions on how could a state defy central welfare legislation. The bench asked the central government to take steps to enforce across the country. This is a human-political problem and nothing to do with the weather.

The two natural resources that are vital for agriculture are the soil and the climate. Man has no control as yet on the climate and needs to adapt to it. The farmer is aware that crop production is limited by climate. He also knows that some crops do well in his region whereas others do not. In tropical regions like India, the production is limited by moisture while in extra-tropical regions it is limited by energy. When a model developed under extra-tropical conditions [SORGF of Texas A&M] tried to apply to semi-arid tropical conditions and found it fails, with poor correlation coefficients both in terms of production of biomass and grain yields. This is even after modifying several sub-models relating to energy factor. When the moisture sub-model was replaced with moisture model developed for the tropical semi-arid conditions the correlation coefficient gone up from 0.35 & 0.37 to 0.85 & 0.81 for biomass and grain yield. Crop development is related to several factors like temperature, photo-period, relative humidity, soil moisture, soil fertility, crop population, etc. Each crop and variety has a range of temperature, where its development needs are met. As a result temperature is not an important factor in India. The main factors for crop production are precipitation/irrigation and fertilizer. Under irrigated agriculture, yield is a function of several factors including soil type, fertilizer, water [quality & quantity], water logging/salinity factors, etc.

The issues that are pertaining to the impact of climate change on agriculture were discussed by **Reddy (2011).** The two biological processes that are influenced by weather parameters are crop development and growth. Development is the appearance of a phase or series of phases during a plant's life cycle; and growth refers to an increase in weight or volume of a certain organ of a plant, or a plant as a whole, within the time interval of a certain phase or an entire life span of a plant. Sometimes these processes are hindered by unusual factors such as insects/pests/diseases on the one hand and frost, hail, floods, strong winds, excess radiation or temperature, cyclones, manmade factors on the other. These, by their intensity may cause death of the species and their frequency renders its cultivation uneconomical. The former type again relates to weather on one hand and the soil on the other. That is, weather influences the degree of interaction between crop and insects/pests/diseases. Some these are discussed by **Reddy 1993.**

Thus crop development is defined principally by temperature and the growth is defined by principally by water, i.e., soil moisture reserve at different growth stages in rainfed agriculture under irrigated agriculture it is not a limiting factor and it is same with soil recharged by winter snowfall. Historically farmers developed crops/cropping systems, known as farming systems with several years of experience for a given soils and climates of the location and region. They have a tolerant range in terms of weather under different soils **[Reddy, et al. [1984b], Reddy, 1993]**. The agriculture scenarios in Andhra Pradesh and in India are discussed in **Reddy [2000, 2007 & 2011]**. The climate implications are discussed in **Reddy [1993, 2002]**.

As the world's population doubled, irrigated farming expanded from 40 million hectares to almost 300 million hectares today a seven fold increase [**Reddy, 2014**]. This revolution in water technology improved crop yields and enabled farmers to grow additional crops each year. China, India, Indonesia, and Pakistan together accounted for almost half the world's irrigated area and thus rely on irrigation for more than half their domestic food production. Global agricultural food production already accounts for 70% of all water withdrawn from rivers and aquifers. However, more than 60% of cultivated area is still at the mercy of Rain God in India and as well in Andhra Pradesh and thus on soil moisture; and yet produces the remaining around 50% of food production. Climate change will only make matters worse in all these situations [**Reddy, 2011**].

River water and to a maximum extent the aquifers are associated with the precipitation in the catchment areas while the soil moisture is based on *in-situ* precipitation. So when we are looking at climate change, in precipitation we must account not only in situ precipitation but also catchment area precipitation and their climate change. Temperature principally refers to *in-situ* temperature only. However, in some cases advection plays important role. The growing conditions, in terms of climate, are largely controlled thermally at higher latitudes and Hydrologically at lower latitudes. It appears that researchers under the former place more emphasis on energy factors and less emphasis on water in crop yield models. Thus models developed under extra-tropical conditions when applied to tropical conditions, it invariably fails. For example, SORGF model developed in USA applied to Indian conditions resulted poor prediction. By replacing soil water balance model the predictions improved from 0.35 to 0.85 in terms of correlation coefficient. The SORGF water balance sub-routine works well under conserved soil moisture conditions but fails under irrigation/rain-fed conditions See **Reddy 1993, 1995a, 1995b**.

This kind of broad range of growing conditions seems common. Pick up any vegetable seed packet from your local garden shop, and you will see instructions for growing the plant in a wide range of climatic conditions, in different regions of your country. Staple crops like Wheat are grown in conditions ranging from Canada and Siberia, on the edge of the Arctic, to the blistering hot Australian Outback. Warmth loving plants such as tomatoes sometimes have to be started in a greenhouse, when grown in cold countries. On a commercial scale, this is often achieved by covering acres of new seedbeds with plastic.

The successful adaption of rice to a wide variety conditions is even more spectacular by selective breeding, providing the world with new, high yielding varieties of rice. Other plants like strawberries thrive in cold England, but they also thrive in subtropical Australia, providing you keep them well watered. Even coffee bushes can cope with a broad range of environmental growing conditions. There are commercial coffee growers in Australia, who produce coffee in climatic conditions far removed from more traditional alpine equatorial coffee producing countries. In my opinion, it is no exaggeration to suggest climate scares about food production in a warmer world are nonsense. It is good to see industry bodies like the International Organization of Vine and Wine starting to push back against such claims.

Dr. M. S. Swaminathan observed "Global warming will affect agricultural productivity adversely in regions where the food needs are the greatest". A one degree rise in global temperature will curtail the growing season by a week in Punjab, he said. "You lose one week of growing period, you lose nearly 400 kg of wheat per hectare. We have calculated 6 million tons per year will go down in the northern part of India. International Food Policy Research Institute has estimated that food production in south Asia will decrease by 44 per cent by 2050, if adaptation measures are not put in place". All these are populist statements by the so called big people. IPCC presented a theoretical exercise using model prediction on the impact on agriculture with the increasing levels of temperature rise [**Figure 4.6**].

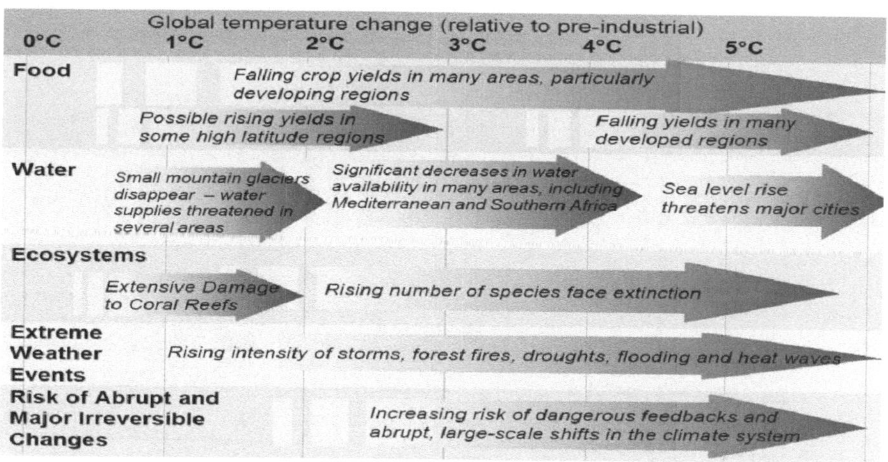


**Note:** though IPCC is used climate change, as it relates to global warming, the title is corrected accordingly

**Figure 4.6** Projected impacts of Global Warming level on Water resources, agriculture & ecosystems as per IPCC

**Table 4.4** presents the average and extreme temperatures of maximum and minimum for Hissar, Ludhiana & New Delhi covering the wheat belt in Punjab – the data was taken from IMD Red Book [Normal of 1931-60]. That means the observed year to year variations are abnormally high as this region is affected by summer heat waves and winter cold waves, whose intensities vary from year to year depending upon regional circulation patterns. Also in wet and dry precipitation years the temperatures also change -- During 2002 and 2009 drought years with 81 and 73% of average, the temperature recoded higher by 0.7 and 0.9 °C [**Figure 4.7**].

**Table 4.4** Temperature variations in three wheat growing zones in India

| Parameter | Temperature, °C | | | | | |
|---|---|---|---|---|---|---|
| | Hissar | | Ludhiana | | New Delhi | |
| | Max. | Min. | Max. | Min. | Max. | Min. |
| Mean | 32.8 | 17.4 | 31.9 | 17.3 | 31.7 | 18.8 |
| Highest – month | 46.3 | 00.8 | 46.4 | 01.4 | 44.9 | 02.7 |
| Highest – day | 48.3 | -3.9 | 48.3 | -1.7 | 47.2 | -0.6 |

Max. = maximum; Min. = minimum

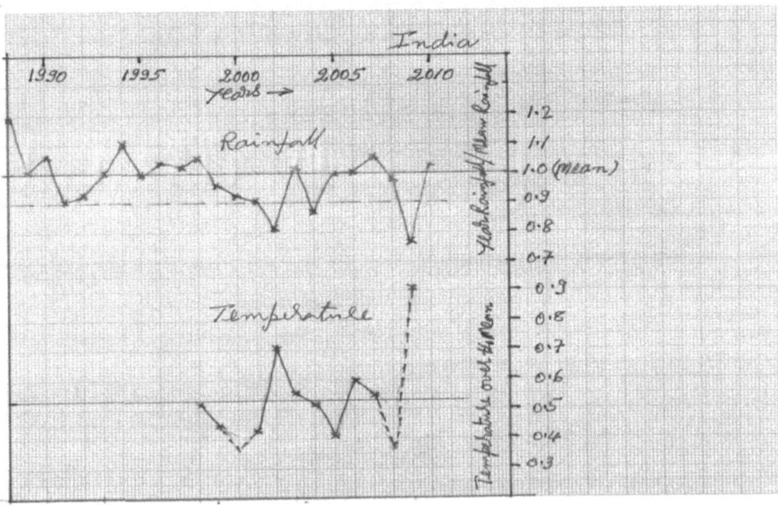


**Figure 4.7** Inter-annual variation of temperature with rainfall

The **Table 4.5** presents the effect of temperature and relative humidity on few sorghum varieties (**Reddy, et al., 1984b**). The equation presents the days taken to maturity of four sorghum varieties. When the temperature coefficient is low the tolerance of that variety to temperature is high, which means that variety can be grown under wider temperature regimes. That means, CSH1 variety has higher temperature tolerance and next comes Local variety. Because of this character CSH1 is grown in both kharif and rabi seasons. M35-1 variety is less tolerant to temperature. It is mostly grown in rabi season. Thus, crop development prediction equations assist in the identification of areas where a given variety is most suitable along with its tolerant limits. In the traditional agriculture, based on the several hundreds of years experience farmers identified crops and cropping systems to suite the soils and climate of the location and region. The new varieties lack this and thus needs to characterization through the above stated type models. The degree day or heat-units concept assumes that constant heat units are required to reach a particular growth stage. Under this condition, when temperatures are higher the growth stage reaches early and when the temperatures are lower it delays.

**Table 4.5** Regression coefficients of predictive equation for sorghum cultivars

| Phenophase | Genotype | Regression parameter | | | |
|---|---|---|---|---|---|
| | | a | b | c | R |
| Days to harvest | CSH1 | 133.2 | -12.6 | -0.35 | 0.83 |
| | Local | 152.5 | -13.9 | -0.39 | 0.67 |
| | V302 | 149.5 | -16.5 | -0.48 | 0.79 |
| | M35-1 | 143.5 | -16.9 | -0.31 | 0.74 |

Equation: $D = a \pm b \times [|19.6 - T|]^{1/3} + c \times RH$,

where D is the days to the harvest, a is the regression constant in days, b is the temperature (T in $^{o}$C and it is more than equal to 19.6) coefficient, c is the relative humidity [RH in %] coefficient and R is the multiple correlation coefficient.

As long as the crop meets the moisture requirement the delay or early reaching the maturity stage will not affect the yield. That is under irrigation the delay or early by few days in the completion of growth will not affect production in locally adapted varieties. In rainfed agriculture, however, the yield reduction relates to water stress in different growth stages. Here any delay in that growth stage further amplifies the stress impact. At certain growth stages the stress has more affect on yield compared to some other stages. This varies with soil type. **Reddy (1993)** presented concepts for crop and cropping pattern adaptation to climate change and soil type under rain-fed agriculture and the role of water stress at different growth stages.

Now let us look at the production figures of sorghum cultivar SPV-462 under different land management conditions (**Reddy, 2000**) during 1989-90, a good rainfall year with 1356 mm and during 1992-93 a poor rainfall year with 525 mm (**Table 4.6**).

**Table 4.6** Yield of sorghum cultivar SPV-462 in good and bad rainfall years under different soil management conditions

| Soil management | Sorghum cultivar SPV-462 Yield, kg/ha | | |
|---|---|---|---|
| Year | 1989-90 | 1992-93 | % |
| Precipitation (mm) | 1356 | 525 | |
| Sowing along the slope | 1923 | 0488 | 25 |
| Sowing across the slope | 2596 | 0743 | 29 |
| Sowing on contour lines | 2692 | 1125 | 42 |
| Sowing on contour lines with dead furrows | 3461 | 1238 | 36 |

Yerracheruvu watershed in Anantapur during 1984-85 to 1990-91, the lowest yield of groundnut was 323 kg/ha recorded in 1989-90 and the highest yield of 1840 kg/ha was recorded in 1986-87. In Andhra Pradesh the production (lakh tons) during 1995-96 to 2002-03 are 116.64, 136.81, 108.22, 149.10, 136.97, 160.27, 148.36, 102.85. The difference between the highest and the lowest is 57.42 lakh tons, around 30% of the average. With the green revolution, technology drought is added to weather drought. That is under chemical inputs, weather drought impact accelerates the stress and thus reduction in yield. That means, at local and regional level production is influenced by several factors and if and only if the climate change limits exceed these weather aberrations then only the yield is affected. Models may tell stories but none of those models take into accounts all such factors.

**Poverty and food security:** Global warming is nothing to do with Poverty & Food security – it relates to bad governance. FAO report states that around 30% of food produced is going as wastage globally – in India it is around 40 to 50%. Recent report says in USA this waste is 40%. New agriculture technology is affecting severely the nutrition security. For more details see **Reddy [2011].**

Falling living standards are contributing to a schocking surge in malnutrition, and 'Victorion' diseases which wereprevalent in the 1800s soaring in England 'due to food poverty and cuts – devastating job losses, electricity costs have surged/the price of gas has also surged'. **Figure 4.8** presents Electricity costs as a function of per capita installed renewable capacity. Wind and solar only, excludes hydropower.

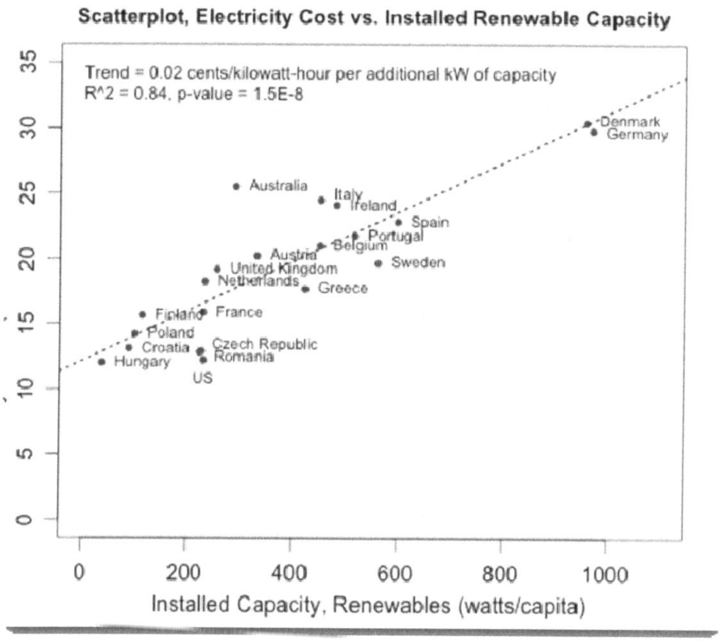


**Figure 4.8** Electricity costs as a function of per capita installed renewable capacity. [Wind and solar only, excludes hydropower.]

**Food price rise:** Prices of the commodities in the market are not related in any way to production level as we always produce in excess and store for deficit years. Even in a good production year prices are not coming down but on the contrary goes up. In good production years large part is exported legally and illegally. Profit driven speculation, a political game, controls the food prices at international markets. Finally the bad governance plays the vital role in price rise also. Global warming has no role in food price rise.

**Food production**: Drastic variations in area and production in AP during the last few years (**Table 4.7**). These variations are in no way associated with global warming phenomenon but they are the manifestation of new agriculture technology. **Figure (4.9a)** presents the yield variation with fertilizer use in Andhra Pradesh and **Figure (4.9b)** presents the same for national level. This figure also includes few other components. **Figure 3.15** presents the drought proneness zones in India and **Figure 3.16** presents the variation of drought proneness with the climate change [fluctuations].

**Table 4.7** Production and area in Andhra Pradesh during 1999-2000 to 2008-2009

| Crop | Area [Lha] | Production [Lt] |
|---|---|---|
| Paddy | 38.83 [28.22-43.87] | 121.42 [73.27-142.10] |
| Oilseeds | 26.75 [22.74-29.22] | |
| Groundnut | 17.22 [13.34-18.76] | 014.62 [08.20-026.04] |
| Cotton | 11.43 [08.03-13.98] | 026.87 [10.86-034.91] |

| Year | G'nut | Cotton | Maize | S'ghum (Lha) | PMillet |
|---|---|---|---|---|---|
| 1984-85 | 16.76 | 05.54 | 03.13 | 18.62 | 03.95 |
| 1990-91 | 23.91 | 06.65 | 03.09 | 11.90 | 02.31 |
| 1995-96 | 22.20 | 10.59 | 03.33 | 08.80 | 01.33 |
| 2008-09 | 17.22 | 11.43 | 08.56 | 02.89 | 00.58 |
| Irrigated | 7.0 | 3.0 | 2.1 | 0.3 | 0.4 |

[% of net area irrigated during 1995-96]

**Role of Chemical Fertilizers in yield**

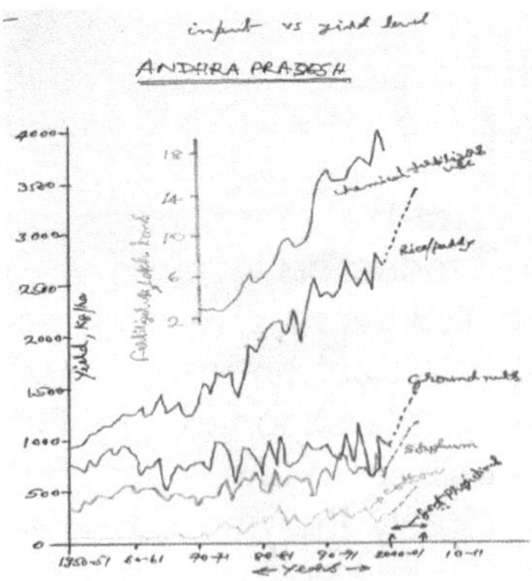


[a]

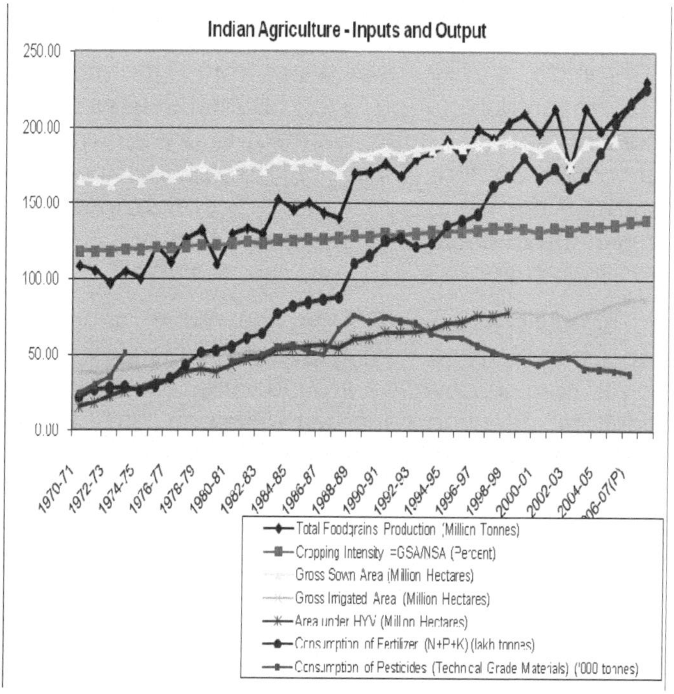


**[b]**

**Figure 4.9** Role of Chemical Fertilizer in Yield [a] Andhra Pradesh & [b] All India

Food includes not only agriculture products but also include: Dairy products, Poultry products, Sea & Water products [fish & prawns], Animal products [meat], etc. To provide economic and nutrient security animal husbandry linked agriculture play important role in traditional agriculture but with chemical input based/GMO based technologies affected this severely. So, technology plays the major role in nutrient and economic security to farming community in specific and rural population in general.

I presented a talk on All India Radio, National Network Programme recorded on 18th February and broadcasted from Delhi on 11th March 2011 at 10 PM on the "Role of climate change on agriculture production, price rise and food-nutrient security". Now it is widely discussed subject by FAO, World Bank, CGIAR & Oxfam – attributing to climate change, which is a misleading notion to serve vested interests. Traditional food is not only nutritious and healthy food and present food is unhealthy, which is a part of PDS.

## 4.5.2 World Bank Prophecies on Agriculture

Deccan Chronicle, a daily News paper from Hyderabad dated 20th June 2013 presented a report "World Bank releases report, paints grim picture of future – Global warming to dry up rivers, inundate cities". It observed that India's monsoon will become highly unpredictable if the world's average temperature rises by 2 °C in the next two-three decades. Deccan Chronicle also presented below the World Bank report my counter on that.

Indian agriculture scientists in the same tone and tenor forecasted using El Nino saying that it is destroying agriculture, production is coming down by around 20%, severe losses to dry-land agriculture, affecting northeast monsoon, etc. **Figure 4.10a & b** respectively presents the annual rainfall march on which marked El Nino and La Nino years; and food grains yield on which marked El Nino years for three intensity categories in India. By looking at yield levels versus El Nino intensity [1–weak, 2–medium and 3–strong], there is no correlation; and so also the case with annual rainfall; and yet they deduced such strange conclusions to create panic among people.

**During below the average 30-year period cycle [1897-1926; 1957-1986] – El Nino presents drought conditions and similar pattern with La Nina and floods is not seen**

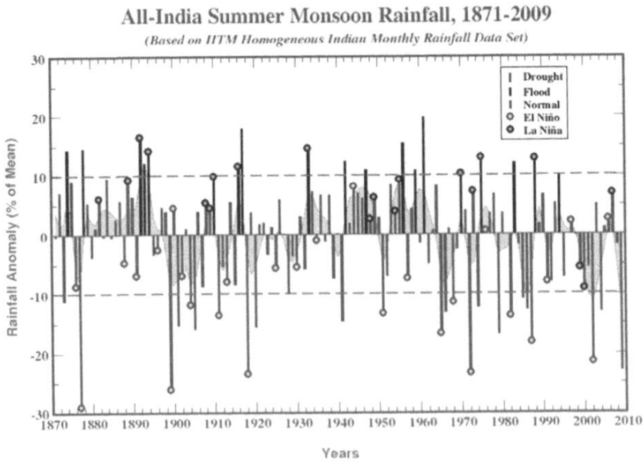


[a]

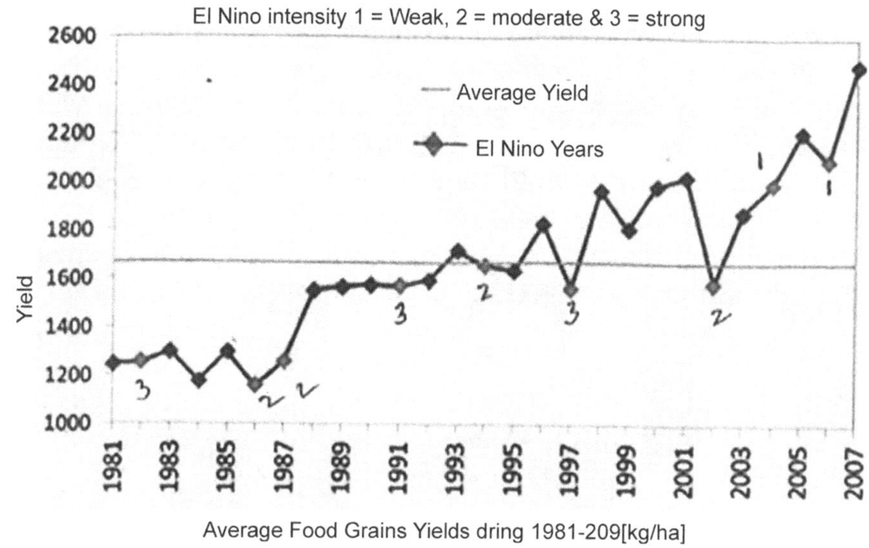


**[b]**

**Figure 4.10** Impact of El Nino on [a] Precipitation & [b] Crop production in India

In fact the yield is a function of chemical inputs, which I published more than a decade back. **Figure 4.11a, b & c** present respectively fertilizer versus yield of paddy, changes in crop area and changes in area under irrigation under different sources. Also, reports after reports were published saying that production is less than the demand and thus price rise and affecting food security. At the same time FAO report presented that globally 30% of what is produced is going as waste. This is 40-50% in India. Now a report states that it is around 40% in USA. That means to that extent the natural resources & finances used to produce that are going as waste. **Table 4.8** presents the subsidies provided by the Indian Government for fertilizer and food. In India, the main reasons for such losses are lack of storage facilities, and lack of timely transportation facilities. All these are nothing to do with global warming but it relates to poor governance and businessmen manipulations.

In India, though planned to link-rivers to improve the quality of life in rainfed areas, vote bank politics creating more problems under the disguise of interstate and intrastate disputes on water sharing [**Reddy, 2015**] on one side and World Bank on the other side. World Bank and its PR groups are working against building dams in India. Now they say climate change as a culprit. Unfortunately, climate change is a part of dam development. Rainfall variations were taken in to account while building dams.

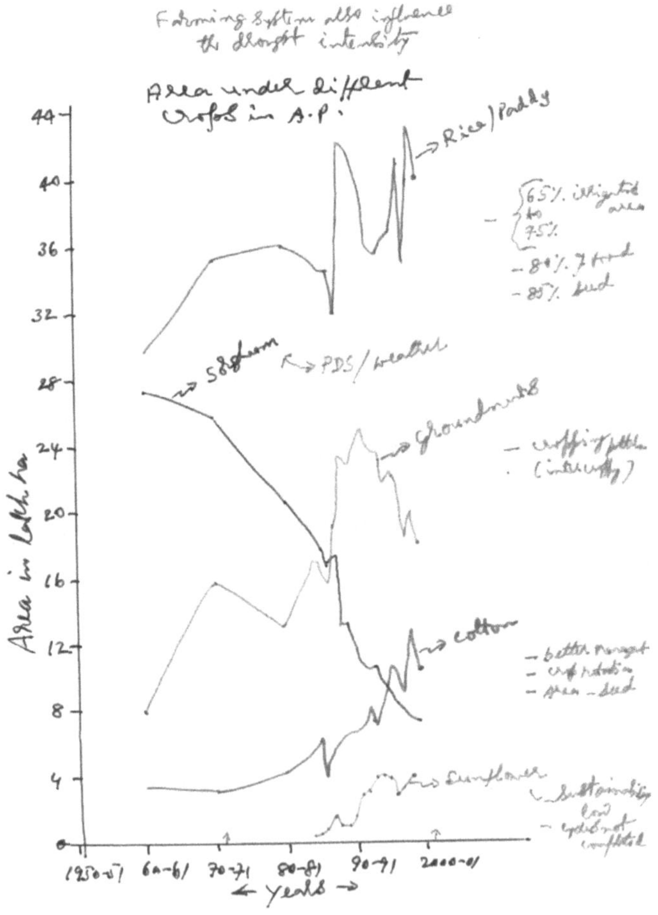


[a]

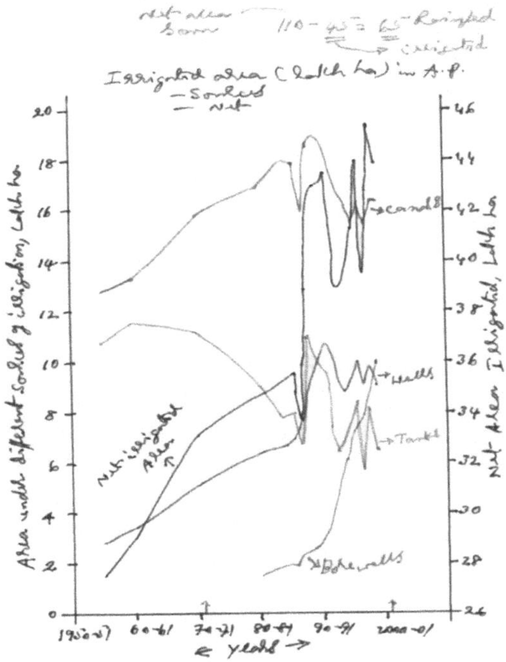


**[b]**

**Figure 4.11** [a] change in crop area; and [b] change in area under different irrigation sources in Andhra Pradesh

**Table 4.8** Fertilizer and food subsidies in India

| Subsidies | 1990-91 [in crores | 2008-09 of Rupees] |
|---|---|---|
| Fertilizer | 4,389 | 75,849 |
| Food | 2,450 | 43,627 |
| | **1990-91 to 2000-01** | **2001-02 to 2008-09** |
| | **[in % of total subsidy]** | |
| Fertilizer | 43.0% | 39.5% |
| Food | 35.1% | 49.1% |

-- Fertilizer subsidy 2006-07 to 2008-09 = 58.7% & 2000-03 = low % -- more than 30% misused in all the subsidies, Fertilizer is diverted to Aquaculture, etc.

## 4.6 Concluding Remarks

In agriculture, weather and soil plays an important role. This is modified by the agriculture technologies. The natural variability in climate, a major component of climate change, play the vital role in agriculture as it defines the droughts & floods and the associated dry and wet conditions. For centuries farmers have developed adaptive systems for such climate change. We rarely see this now. Under the new agriculture technology it fails as the proponents of new technology haven't proposed climate related adaptive procedures as the research is limited to few years while the traditional farmers' adaptive measures were based on Centuries experience.

The UN agencies and others look at global warming as the major culprit. It became ritual for several groups even without reading the IPCC's Reports in their totality build stories that create confusion in developing countries. It is a bad practice. Also, it has become a common scenario that people use the word climate change, mostly as an adjective to divert the real problem. Nowadays the media is filling pages with such rhetoric stories; as such reports get hype to the readership.

Most of the inferences that are seen in literature on the impacts of climate change were based on the IPCC model predictions of high temperature trend 3 to 4 °C increase by 2100 but in reality observed data presents only an increase of around 0.15 °C. Therefore to understand the impacts in correct perspective, there is a need to characterize climate change components at local, regional and national level in addition to global level [if any] based on observed data series. Thus, the risk of impacts of climate change on "water resources, agriculture & ecosystems -- W-A-E" could be minimized by developing procedures for adaptation **[Reddy, 1993]**. In nature there are several other factors that impact W-A-E in addition to climate change – such as deforestation; air, water & soil pollution; natural disasters; tourism/pilgrimage, mining, real-estate activities in forests & hilly terrains; agriculture technology after 1960s, etc. However, there is no mechanism to isolate the effects of climate change from these.

Where the climate data presents cyclic variations, the truncated data of such cyclic variation gives spurious conclusions, which are dangerous. Finally, therefore, there is an urgent need to revive research mode of 70s & 80s, which is unpolluted by global warming, by research institutions instead spending time & money on IPCC mode time pass research. Ignorance is terrible but exaggeration is dangerous.

If we look at the daily weather charts for the past 100 years and the newspaper reports during the same period, it is clear that the extreme weather events what we are experiencing today are not new but they were there in the past even with more severity. It is the right time that UN agencies instead of harping on global warming and carbon credits and spending billions/trillions of dollars, should look in to collecting these extreme weather events over different parts of the globe, year by year.

5 Chapter

# Summary and Conclusions

***"Global Warming is a science of building castles in the air; and Climate Change is the science of weather variations with space and time that allows development of sustainable adaptive measures"***

***We must remember the fact that the Earth's climate is dynamic and always changing through a natural cycle. What we are experiencing now is part of this system.***

**Reviews of climate change Reconsidered II: Physical Science:** *"The CCR-II Report correctly explains that most of the reports on global warming and its impacts on sea-level rise, ice melts, glaciers retreat, impact on crop production, extreme weather events, rainfall changes, etc have not properly considered factors such as physical impacts of human activities, natural variability in climate, lopsided models used in the prediction of production estimates, etc. There is a need to look into these phenomena at local and regional scales before sensationalization of global warming – related studies" by S. Jeevananda Reddy, Formerly Chief Technical Advisor United Nations World Meteorological Organization.---*

Open Letter to the Secretary-General of the United Nations [29 November, 2012] – I am one of the signatory of the letter

***"Policy actions that aim to reduce $CO_2$ emissions are unlikely to influence future climate. Policies need to focus on preparation for, and adaptation to, all dangerous climatic events, however caused"***

## 5.1 Climate to Climate Change to Climate System

**Weather and Climate:** Climate is what you expect; weather is what you get. Weather refers to short-term events in the atmosphere, while climate relates to longer time period events in the atmosphere. Every National Meteorological Services published Climate Normal books [in India, IMD (India Meteorological Department) called this as Red Book] that contain meteorological station-wise averages and extremes for individual meteorological parameters such as temperature, precipitation, wind, humidity, etc. The period used was for 30 years [1931-60, 1961-90, etc]. Thus, temperature is only one parameter of weather and climate. The Earth's temperature has been stabilized at 15 $^{o}$C over the centuries before the industrialization through the interaction of Sun's Energy with the natural greenhouse gases in the atmosphere under greenhouse effect. However, this presents intra-annual and intra-seasonal, intra-regional, intra-climate system, etc., variations which are of importance for adaptation. These are modified by the general circulation patterns.

**Climate Change:** The word climate change was in use even in 17$^{th}$ century. This is primarily relating to agriculture scenario, which means the systematic variations in climate, more particularly in precipitation that defines the droughts and floods. Moisture is the limiting factor in tropical warm countries wherein most of the developing countries are located; and thus knowledge of natural rhythm in precipitation, a major component of climate change, played the pivotal role in water resources availability assessment; and thus success and failure of agriculture. Our forefathers adapted to this for centuries and yet UN agencies haven't given importance to such valuable experiences. This played the major role in the reduction of poverty, which is the primary objective of Pope Francis Encyclical of 2015. However, new agriculture technologies that don't fit in to weather-climate system provided scope to misrepresent success or failures to climate change.

Within the past three decades, IPCC & UN Agencies changed the definition of climate change and thus created confusion among readers in a variety of ways in which media played spoilsport. Also, they introduced new words in place of global warming such as anthropogenic climate change, human induced climate change, anthropogenic global warming, and human-induced global warming. A simple science was made in to a complicated issue. Thus, the science of "climate change" has turned into a political satire of "global warming and carbon credits". Also, majority use climate change as an adjective to get their papers published in reputed journals or institutions to get funds or to go with the wave. IPCC is giving weight to number rather than science. In this process the major casualty is the health of life forms on the Earth, more particularly in developing countries. The UN system now is shy of using the very word global warming and thus creating confusion among people, leaders, administrators, scientific journals, scientific institutions and finally policy makers.

With the passing of time the UN agencies [IPCC's AR3 to AR4 to AR5] created confusion on the definition of climate change as they were confused on the definition of global warming. From these confusions, we can see that "climate change refers to a statistically significant variation in either the mean state of the climate or in its variability, persisting for an extended period (typically decades or longer). Climate change may be due to natural internal processes or external factors such as persistent changes to the atmosphere or changes in land use. That is a change of climate which is attributed directly or indirectly to human activity that alters the composition of the global atmosphere and which is in addition to natural climate variability observed over comparable time periods. People mainly perceive climate change as changes in atmospheric variables, for example, variations in temperature or precipitation. In principle, due to its chaotic dynamics, the atmosphere itself can generate many natural climatic changes fluctuations."

From all these confusions, climate change can be defined in simple way and easy to comprehend is as: Climate change consists of two parts, namely (1) systematic variations expressed by cyclic variations and (2) human-induced changes expressed by trend. Former is natural inbuilt variations further divided in to 1(a) irregular variations associated with the intra-seasonal and intra-annual variations; and 1(b) fluctuations or rhythmic variations or cyclic variations. These are beyond human control and thus needs to adapt to them. That is exactly what our forefathers did in the case of water resources and agriculture. The later is further divided in to two parts, namely 2(a) associated with greenhouse effect; and 2(b) associated with non-greenhouse effect. The non-greenhouse effect is known as "Ecological Changes" associated with changes in land & water use and land & water cover.

Thus, the global average temperature anomaly consists of two parts, wherein the natural fluctuation is superposed on trend. Global average [surface + oceans] temperature anomaly "trend" is contributed by two components, namely 2 (a) and 2 (b). IPCC in its

AR5 observed that the contribution of 2(a) is more than half and 2(b)'s contribution is less than half [a qualitative statement, which has no meaning in science]. 2(a) is further sub-divided in to 2(a)(i) contribution by anthropogenic greenhouse gases since 1951, a human induced effect; and 2(b)(ii) contribution by other anthropogenic forcings associated with non-human activities such as volcano eruptions, etc, a natural effect. IPCC observed that the former provides warming and the later provides cooling; but it is not so as in addition to cooling, it reduces the incoming energy that is used in greenhouse effect. 2(b)(i) is termed as "Global Warming" wherein anthropogenic greenhouse gases related as people burn fossil fuels. 1(a & b), 2(a)(ii) & 2(b) are location-region specific activities. With reference to anthropogenic greenhouse gases [2(b)(i)], some are location-region specific and with the shorter lifespan like Methane; and some like carbon dioxide is though also location-region specific used as global [they are presuming that with the wind they mix] with longer life – but now reports say only 40 years. Though the Methane has a 70 times more powerful over the Carbon Dioxide and yet reports state that its contribution is relatively small.

Through the greenhouse effect it changes energy in to temperature. Atmosphere contains large quantity of greenhouse gases, it is not clear, over and above this the new additions are going to convert energy in to temperature through greenhouse effect, if so how much, is a big question as because the Earth life is sustained over millions of years. It is a fact that the energy to be used has a cap and it is not unlimited. Thus, this has reached a saturation point and thus any addition of greenhouse gases to atmosphere will not change the temperature any further. Report states that "All the light emitted by the surface in the strongest $CO_2$ bands was completely absorbed at pre-industrial levels. That is below 280 ppm."

Ecological changes are not part of greenhouse effect though they are part of human created changes. They present uneven affects based on the intensity and spatial spread. Because of this they present high regional variations around the globe and yet it is a continuous process part of trend. With the human civilization, this also changed. IPCC used 1951 as the starting year for global warming and yet we can see the change in global average temperature anomaly starting from even before 1880 from the time instrumental data are available. Here we must remember one fact that fluctuations of centuries present a component, a linear trend [increasing or decreasing] over shorter periods but they are insignificant. See some reports, what they say:

Like greenhouse gases and global warming, when the term climate change is used, everyone now assumes the human-induced variety is being discussed when, in fact, the speaker or author might be referring to naturally caused variations in climate. Most of the times IPCC and other government agencies used *climate change* for both natural and human-induced variability often without identifying which they're discussing. It gives an impression: you assume they were discussing the man-made type, when in fact they may have been referring to natural variability. Some persons might believe the IPCC and other governmental agencies are purposely being vague, with hope that most readers will assume they're discussing the man-made kind, when in reality those scientists still can't differentiate between natural and anthropogenic climate change. In essence, climate variability looks at changes that occur within smaller timeframes, such as a month, a season or a year [known as irregular variations intra-seasonal, intra-annual variations], and climate change considers changes that occur over a longer period of time, typically over decades or longer.

A key difference between climate variability and change is in persistence of "anomalous" conditions when events that used to be rare occur more frequently, or vice-versa. In statistical terms the curve of the frequency distribution representing the probability of specific meteorological events changes. The curve may be modified either in amplitude, shifted about a new mean, or both. Care must be taken not to confuse variability with change. Many regions of the world experience greater variability, climatologically speaking, than do others. In some parts of the world, or in any region for certain time periods or parts of the year, the variability can be weak (i.e., there is not much difference in the conditions within that time period). In other places or time periods, the conditions can swing across a large range, from freezing to very warm, or from very wet to very dry and exhibit strong variability. A certain amount of this is understood and accepted, instinctively, by the people in a region.

What is "normal" for Denver, Colorado in terms of the frequency of precipitation events (high variability) would be "abnormal" for

London, England (low variability). Thus, any single event, such as a severe tropical cyclone, cannot be attributed to human-induced climate change. Occasionally, an event or sequence of events occurs that has never been witnessed before (or recorded before), such as the exceptional hurricane season in the Atlantic in 2005. Yet even that could be part of natural climate variability. If such a season does not recur within the next 30 years, we would look back and call it an exceptional year, but not a harbinger of change. Only a persistent series of unusual events taken in the context of regional climate parameters can suggest a potential change in climate has occurred.

The atmosphere itself can generate many natural climatic changes such as fluctuations like the Northern Atlantic Oscillation (NAO), which significantly influences the climate over parts of Europe and North America. It is a kind of pressure fluctuation between the Iceland Low and the Azores High that determines the strength of winter westerly winds across the North Atlantic. If these are strong, the result is mild and rainy weather in Western Europe; if they are weak it is dry and cold.

**Climate system:** The climate system consists of five major components, namely the atmosphere, the hydrosphere, the cryosphere, land surface, the biosphere. The climate system is continually changing due to the interactions between the components as well as external factors such as volcanic eruptions or solar variations and human-induced factors such as changes to the atmosphere and changes in land use & land cover and water use & water cover, known as ecological changes. General circulation patterns are superposed on the climate system. These two systems play vital role at local and regional weather and climate.

## 5.2 Reliability of Global Average Temperature Anomaly

To validate all the theories or hypotheses that relate to global warming & climate change, we need accurate data of weather, more particularly temperature; and greenhouse gases, more particularly carbon dioxide; in both space and time that covers climate system and general circulation patterns.

At present we have temperature data measured directly and indirectly. Direct measurements have been made in meteorological stations established as per the procedures laid down. But later on there have been several modifications in terms of unit of measurements, instrumentation, met station design and location, etc. These are started around 16$^{th}$ century but we have global average temperature data series from around 1880. Indirect measurements first started using balloons and later using satellites. Satellite data measurements though started around 1972, the data was available from 1979. The balloon data is available for around 20 years before the satellites data availability. Satellite data covers the globe covering climate system and general circulation patterns, land use and water use changes & cover including the filth cover over land and oceans. Balloon data supplements the satellite data. In the ocean areas, ships used to record the temperature on its way. From around 1990 established a network of buoys.

Though oceans cover 70% of the globe the surface network is relatively very sparse. Thus, the ground based data measurements covered around 20 to 25% of the globe only and that too this level has been reached from sparse network. They are not covering climate system and general circulation patterns over the globe. The global average temperature anomaly was built on poor quality & from sparse & unevenly distributed met network measurements on the Land and in the Oceans/Seas in space and time. The institutions who are maintaining data bases every now and then are changing the data.

**Figure 5.1** presents a classical example of NINO3.4 zone temperature observations starting from January 1870 to October 2015. The MINO3.4 region covers 5°S – 5°N in between 170°W 120°W with an area of 6.2 million square kilometers. Prior to 1990 surface temperature depended on ships traffic. Since 1990s, NOAA has had moored buoys. 20 buoys are in MINO3.4 region. Under General Circulation pattern, 20 observational sites in an area of 6.2 million square kilometers will not give a realistic temperature of that region. Same is the case with surface temperature measurements and greenhouse gases measurements with minor degree of change. We are trying to build castles of global warming based on such a poor network data series. We find hundreds and thousands of papers on the trend month by month; and playing game with the data.

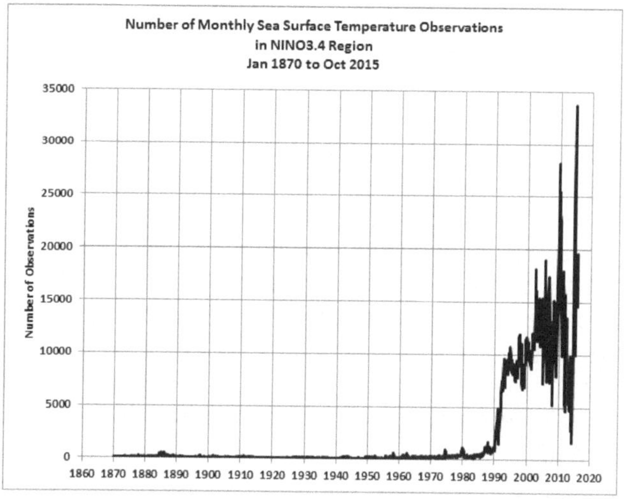


**Figure 5.1** Number of monthly sea surface temperature observations in NINO3.4 region [January 1870 to October 2015]

Since around 1972 [effectively since 1979] satellite data measurements were carried out. The compiled satellite data was released and later withdrawn from the internet as this data series was lower than the ground based global average temperature anomaly by about 0.3 °C and thus it does not fit into their theory of global warming prophecy of doom. Model estimates were linked to ground based data series, which were built by lowering the past data and raising the current data series. If we look at US average raw temperature anomaly or PDO, they clearly show a 60-year cycle with little or no trend.

A report state that "The differences between Sea Surface Temperature data sets prepared by different groups prevent from knowing which El Nino was strongest according to NINO3.4 Region Temperature data." The same is the case with identifying drought affected regions according to India Meteorological Department data [observed versus normal] and State Governments data [observed versus normal]. Here the problem is selection of stations and numbers of stations used to get the index play vital role for the differences. It is the reality of climate system and as well with general circulation patterns. Global temperature anomaly curve derived by different groups also differ but all built the data by manipulating time and again. Since 1997/98 El Nino, the manipulation is severe in nature and particularly the ocean temperature has been raised steeply. During this period, the satellite data showed zero trend but surface data showed a rise by about 0.25 °C.

For these areas where measurements were not available or sparse they were filled using grid network, data was interpolated and extrapolated in a climate system with wide variations. Same is the case of carbon dioxide observations. Satellite and balloon data [balloon data is available even beyond 1979] from 1979 presents a different picture over the ground based data. The North Polar Region has not warmed for about 14 years [UAH], the globe has not warmed for 22 years [UAH], & the South Polar Region has not warmed for 37 years [UAH]. In fact no amplification of warming has presented itself, ever since industrialization, except when ocean cycles are in their warming regimes. No long term warming until El Nino has happened for 18 years and 9 months [RSS].

Let us see few aspects of average temperature -- ground based measurements versus satellites based measurements: (a) traditionally in ground based measurements of temperature, the sensors in maximum and minimum thermometers are mercury and alcohol. Nowadays different sensors are in use. Similarly in satellites and balloons the medium of sensors are different but they are calibrated with the thermometer observations; (b) Ground based average temperature is derived from the averaging of maximum and minimum, thinking it follows a sine curve with the duration of day and night are the same. In reality it is not so. If you take the hourly average data from the thermograph, the day average may or may not be equal to the average of maximum plus minimum based on the degree of skewness in the 24-hourly thermograph data from the sine curve that vary with the cloud conditions and duration of the day and the night. While calculating averages, we take up to 2$^{nd}$ place of decimal. Based on the 1$^{st}$ place decimal value is odd or even, we rounded the second decimal place value. For example: 30.45 are taken as 30.5 and 30.55 is taken as 30.5. If the second decimal value is less than 0.05 then the rounded value is 0.0 and if it is more than 0.05, the rounded value is 0.1 added to first place of decimal value. For example: 30.44 are 30.4 and 30.46 are 30.5; and (c) The globe is covered by climate system/general circulation patterns that vary widely from place to place and region to region this creates complications while calculating averages or extrapolation/

interpolation. Also, with the population growth the land use and land cover and water use & water cover presented wide variations. Also, after the industrialization the filth has been covering land, water & atmosphere. All these play vital role on temperature of a place. These are rarely accounted in ground based data as met observatories are maintained clean; and thus present over emphasis of urban heat island effect on the averages. Also, met network hardly covers 20 to 25% of the area of the globe, more particularly over oceans. Surface measurements were made in °F up to 1956 and there onwards they are in °C.

On land majority of the met stations were located in urban areas with rapid spatial and vertical expansion affecting the land surface by converting it in to concrete jungle and filling with filth and destruction of greenery and water bodies along with the increased levels of pollution. Met network is sparse in rural areas even though rapidly the environment is changing from rainfed dry-land to irrigated wetland agriculture. Thus rural areas are covering the land with greenery/water for more time in a year. All these influence the weather and climate around those locations-regions. Thus, the major contamination of ground based observations on land is the positive bias in urban data and negative bias in the rural data.

In oceans, even after 1990 though buoys built the network but this is not going to account the filth effect over sea surface as the network of buoys is limited this also affect the temperature in deeper layers of the ocean waters. The spread of filth over land and oceans is non-linearly increasing with the time, which comes under ecological changes [under land and water cover] but it also influence greenhouse effect. All these are negated in the satellite data as they cover climate system and general circulation patterns on the one side and urban and rural changes, filth cover, etc., that are taking place with the progression of time. Weather satellites provide wind patterns over oceans, an important component of general circulation patterns that relates to advection part.

Groups are selecting certain period data and try to show abnormal rise in global average temperature anomaly to match the model based predictions. Let us see **Figure 5.2a & b:** This figure presents the 5-year moving average march of US ground based average temperature anomaly data series raw, TOBs-adjusted & homogenized. In **Figure 5.2a** presented 5-year moving average satellite data. UAH & RSS; and in **Figure 5.2b** presented the UAHv5.6 [USA48] data do barely exceed the pre-1960 natural variability box and track close to the TOBs-adjusted temperature, but well below the broadly homogenized temperatures. UAH data series was raised up by 0.6 °C and also shifted ahead of few months to match the peaks and troughs. The anomaly in general presents a clear cut natural variation 60-year cycle.

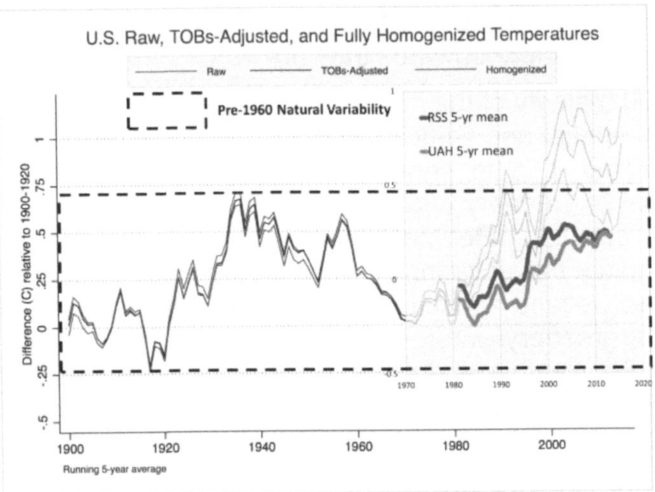


**[a]**

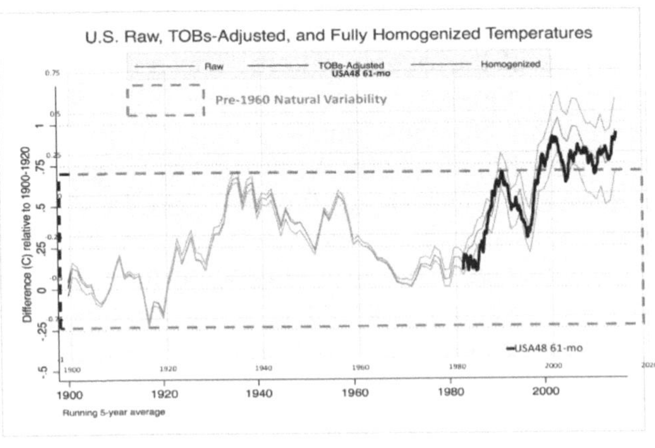


**[b]**

**Figure 5.2** US Annual temperature anomaly march of ground based raw, TOBs-adjusted & homogenized and [a] satellite based UAH & RSS data series and [b] UAH raised and shifted

Though IPCC used 1951 as starting year for global warming, the raw data and adjusted & homogenized data presented no change up to 1970 [even up to 1990] though major share of anthropogenic greenhouse gases were pumped in to the atmosphere by US. From 1990 onwards the differences increased steadily. The 5-year average data series from the two satellites [UAH & RSS] followed the natural cycle. However, some argue that as the satellite data includes the average of multiple layers in the ground to tropopause, the satellite data may not be represented ground data, it will be always lower than the ground data – even if we accept this in terms of magnitude of anomaly, the basic point of interest is presentation of trend, which is far lower than the ground data. The satellite data is generally calibrated with the ground data from selected good stations raw data. A report noted that "In summer ground based data are warmer than upper layers and in winter it is vice-versa". That mean on an average they must present the same.

David Middleton [January 19, 2016] observed "Man-made heat put in oceans has doubled since 1997". McCourtney observed "Why did the heat that was warming the atmosphere suddenly start warming the oceans instead?" Why did ground temperature was corrected from the same period to show there is no pause but rising? It takes about 720 cubic km/yr to raise the ocean level 2 mm/year but the sea level rise show a steady increase of 1.7-1.8 mm/yr starting even before the industrial era. Here we must not forget the fact that the circulation patterns in ocean waters [top to bottom presenting opposite direct] mixes itself and thus the temperature of different layers.

Therefore, the major issue that IPCC and UN must address is to build reliable met data in both space and time instead of harping on "global warming and carbon credits". Here, unfortunately, scientific ethics have gone with the wind. To meet their vested interests groups are mutilating the data collected by different National Meteorological Departments around the globe. Even they are not leaving the satellite data. It is better to close down the IPCC and allow WMO to function in the data domain as it was done earlier in collaboration with national meteorological departments.

## 5.3 Global Warming versus Ecological Changes

Earth's temperature begins with the Sun. Roughly 30 percent of incoming sunlight is reflected back in to space by bright surfaces like clouds and ice. Of the remaining 70%, most is absorbed by the land, ocean, and atmosphere. The absorbed solar energy heats out planet. A report noted that "As the rocks, the air, and the seas warm, they radiate heat energy [thermal infrared radiation]. From the surface, this energy travels in to the atmosphere where much of it is absorbed by water vapour and long-lived greenhouse gases such as carbon dioxide and methane. When they absorb the energy radiating from the Earth's surface, microscopic water or greenhouse gas molecule turn into tiny heaters like the bricks in a fireplace, they radiate heat even after the fire goes out. They radiate in all directions. The energy that radiates back toward the Earth heats both the lower atmosphere and the surface, enhancing the heating they get from direct sunlight. This absorption and radiation of heat by the atmosphere the natural greenhouse effect is beneficial for life on the Earth. If there were no greenhouse effect, the Earth's average surface temperature would be a very chilly –18 °C instead of the comfortable 15 °C that it is today."

What has warmists concerned now is that over the past 250 years, humans have been artificially raising the concentration of greenhouse gases in to the atmosphere at an ever-increasing rate, mostly by burning fossil fuels, but also from cutting down carbon-absorption by forests. Since the Industrial Revolution began in about 1750, carbon dioxide levels have increased nearly 38% as of 2009 and methane levels have increased 148%. However, there is a big question: how accurate these estimates are as the data collection before 1960's was very sparse in space and time; and even after 1960's it is not uniformly spread in space and time? This is very important as the local energy interacts with the local greenhouse gases [natural & anthropogenic] concentrations that result in temperature condition. Global average of greenhouse gases has no role at local or regional level. However, these temperatures are modified by several localized and regionalized conditions including advection.

A report noted that "The atmosphere today contains more greenhouse gas molecules, so more of the infrared energy emitted by the surface ends up being absorbed by the atmosphere. Since some of the extra energy from the warmer atmosphere radiates back down to the surface, the Earth's surface temperature rises. By increasing the concentration of greenhouse gases, we are making the Earth's atmosphere a more efficient greenhouse." But here the basic question is: whether there is still energy left to interact with the additional greenhouse gases?

IPCC, states that "Global Warming" is the increase in the average temperature of global surface air and oceans since about 1950, and to continuing increases in those temperatures. In the last 19 years there is no such increase in global temperature. IPCC [AR4 – 2007] states that "confirms that global warming is now unequivocal and states with more than 90% certainty that increasing concentrations of greenhouse gases produced by human activity very likely has been the primary cause of rising temperatures worldwide since 1950". IPCC [AR5 – 2013] changed "90% and very likely" to "95-100% and extremely likely". It also states that human influence has been detected in warming of the atmosphere and the ocean, in changes in the global water cycle, in reductions in snow and ice, in global mean sea level rise, and in changes in some climate extremes. "It is extremely likely that more than half of the observed increase in global average surface temperature from 1951 to 2010 was caused by the anthropogenic increase in greenhouse gas concentrations and other anthropogenic forcings together". This is highly ambiguous & hazardous statement which allows individuals to think and attribute their own way. The so-called impacts are location-region specific issues. Global average has nothing to do with these changes as the local-regional weather conditions are far from the global average.

The other major component of human activity that disturb the met observations significantly is the ecological changes due to changes in land use & land cover and water use & water cover. These are highly location-region specific. This goes parallel to greenhouse gases increase in the atmosphere with the human activity. However, this has a direct impact on meteorological parameters and indirectly through advection. This also contributes to changes in lower troposphere. Though the ecological changes contribute to global temperature in averaging process but their impact is confined to local and regional issues only. They don't play any role in the temperature impact on nature at global scale.

Urban-heat-island effect and rural-cold-island effect comes under ecological changes. Here heat-island effect may be over emphasized in the averaging of global temperature due to dense network of met stations in urban areas; and less emphasized rural-cold-island effect due to sparse network of met stations in rural areas. Thus, net result of ecological changes is over emphasizing its contribution in global average temperature. The disparity is growing with growth of population-urbanization and wet-land agriculture scenarios world over. Because of this, the Southern Hemisphere temperature is lower than that of Northern Hemisphere. Though ecological changes are part of human activity, they are not part of greenhouse effect. Except global warming component, all other components of global average temperature influence not only temperature but also other met parameters at local and regional scales based on its intensity.

In the global [land & ocean] temperature anomaly data series of 1880 to 2010 following the WMO 1966 methodology separated trend from cyclic variation. From the trend line it is clear that by 2100, the global [land & water] temperature may rise to 1.30 °C from 1880 base year. That is in 220 years the rise is expected as 1.30 °C; this is around 0.6 °C per century. This consists of three components. One of them is emission component. Even if we take this as 50%, then the emission component [global warming] is 0.3 °C per century. Very recently [1-1-2015] British Royal Society and US National Academy of Sciences brought out an overview "Climate change: Evidence & causes". The report included a figure of annual march of global temperature anomaly along with 10 year, 30 year & 60 year moving average patterns using 1850 to 2010 data series. WMO presented a manual on "Climate change" in 1966; and in this manual presented moving average technique as one of the technique to separate trend from rhythmic/systematic variations. Here, the 60-year moving average pattern showed the trend after eliminating 60-year rhythmic variation. The 60-year cyclic pattern was noted by me and the same was included in my books prior to 2010, wherein the sine curve varied between – 0.3 to + 0.3 °C. The trend showed less than 0.5 °C raise during 1950 to 2010. This consists of several components including global warming component.

After correcting this figure for overestimation and underestimation factors discussed above, the trend will be far less than 0.5 °C only. The satellite data is in line with this. From the satellite balloon data this is around 0.25 to 0.30 °C. That is, global warming that has taken place during 1951 to date will be less than 0.125 °C. To reach 2 °C, the time taken surface data with 0.3 °C will be around 650 years; and the same under satellite and balloon data with less than 0.15 °C per century will take more than 1300 years at the present rate of emissions. A report noted that 1990-1950 the rise is 0.45 °C; 1950-2000 is 0.50 °C & 2000-2015 is 0.0 °C, and thus with this plateau condition, they more take more and more time to reach 2 °C.

Here another important issue is the differences in the temperature trends in the Northern Hemisphere [NH] and in the Southern Hemisphere [SH] on the one hand and in the Land Areas and in the Ocean Areas on the other. Though there is a difference in relative activity over ocean and land surface, the regional differences suggests a different story. In the trend, the major component is ecological change. Ecological changes are more in NH and less in SH and zero over ocean we cannot say zero as the shipping activity in oceans is gradually increasing, filth covering ocean waters is increasing oil spills, coastal pollution, plastic waste, etc increasing and thus change the energy balance over the oceans; and also the interaction of ocean waters with the ships movement [friction], water wave movement and their discharges. Majority of ground based data that goes in to the averaging of global average temperature is ecological change part only as majority of the met stations are located in urban areas around the world. Because of this SH temperature pattern is following ocean temperature pattern and NH temperature pattern is following land temperature pattern in terms of its magnitude of changes.

All these suggest there is no global warming associated with anthropogenic greenhouse gases concentration increase in the atmosphere as the energy versus $CO_2$ reached a saturation point with the natural $CO_2$ itself. Also, a recent study reports that $CO_2$ is removed from the atmosphere by natural sinks at a rate proportional to the surplus $CO_2$ accumulation. The half life of the surplus $CO_2$ concentration is approximately 40 years [not 1000 years]. The anthropogenic greenhouse gases time curve is built on faulty data.

Some argued that the term "Global" warming, does not do justice to what is really happening globally. Here is an example of averages of Jan to Dec land temperature anomalies for 2005 -

2014 [last 10 years] – data per NOAA climate at a glance [units in °C/decade]: Contiguous US (-0.68); North America (-0.41); Asia (-0.31); Europe (+0.39); Northern Hemisphere (-0.04) flat; Oceania (+0.07) flat; Africa (+0.08) flat; South America (+0.24); Southern Hemisphere (+0.06) flat; Global (-0.01) flat. So, global warming is not really warming at global scale but it is sum and total of what is happening at local and regional levels only. The alarmists claim global warming is happening at an unprecedented rate. Where people live it is an entirely different picture. Some areas are cooling some are warming and some areas are not changing at all or flat. So a single "global" warming figure is a misrepresentation of what is really happening over different parts of the globe and thus their respective impacts on nature around them.

Canada has no global warming associated with the changes in the anthropogenic greenhouse gases in the atmosphere. The annual mean temperature departures [anomaly] from 1961 to 1990 normal over different regions in Canada for the period 1948 to 2009 presents high variations in rise in temperature [linear trend] they varied between 0.25 to 1.5 °C [www.ec.gc.ca/adsc-cmda/default.asp?lang=En&n=F3D25729-1]. Some areas like Northeastern Forests, Atlantic Canada, Greater lakes and St. Lawrence area show little rise; and some other regions show high rise in temperature. Around the last decade there are sudden jumps and also inter-annual variability show very high in last half period. All these clearly show that this is nothing to do with global warming but it is associated with the ecological changes. Though in 2014 it showed wet regions with colder temperature and dry regions with warmer temperatures [similarly in USA presented east to west vertical divide with moist cold and dry warm; also in India, the dry years 2002 & 2009 presented warm by 0.7 and 0.9 °C at all India level but this was not reflected in the Canada temperature anomaly annual march. Though the later part clearly showed above the average rainfall pattern [longer growing season], the temperatures were also present warmer conditions with high fluctuations. British Colombia presents high wildfires. Though the Canada report on climate state: Warming over the 20$^{th}$ Century is undisputable and largely due to human activities. They did not specify which human activity!

## 5.4 Human Greed

With the population growth and to meet their needs under the modern lifestyle conditions under innovative technologies, the natural disasters impact on environment started increasing. This is not because the severity of disaster has increased but human greed created "increased severity" similar to technology induced drought in agriculture.

Let me give few examples based on my experience from India. In Hyderabad City, where I live, used to have 932 lakes, majority of them were chained lakes, finally the water joins the River Musi through drains/channels. With the passing of time around 500 lakes were filled and converted in to concrete jungle with little infiltration. Even the remaining lakes were filled partially and reducing the water holding capacity. The channels that carry excess rain water to Musi River were occupied by around 90%. Also, sewage is an addition to rain that increases the load with the passing of times. This increases several times flood disaster impact. The year 2000 [August] is a classical example. Water went up to 2$^{nd}$ floor in Ashok Nagar area. Even in my Colony [ICRISAT Colony, Phase-I] in Bowenpally, people were rescued by Military personal through boats as the lakes in this area were completely filled except one and that too around 25% left. In 2015 November due to Northeast Monsoon rains affected areas around Nellore in Andhra Pradesh and Chennai in Tamil Nadu. This is basically because of encroachment of basic drains/rivers that carry rainwater and tanks that store rain water. In addition to rain water sewage is also added to this which increased with progression of time. The roads were laid and buildings were constructed on such drains/river banks. In fact this area used to get more than that of 2015 rainfall per day. Thus, unusual flood disaster is basically caused due to unplanned growth and development in urban areas. Same is the story with urban areas all over India.

During 2009 October floods to Krishna River, Kurnool & Raghavendra Swami Matt in Andhra Pradesh in India were submerged basically because of the negligence from the government by not releasing the water from Srisailam Dam in time, though the Nagarjuna Sagar Dam in the downstream was empty by that time. Here the timely action would have saved these disasters. In the case of 2013 floods in Uttarkhand and 2014 floods in Jammu & Kashmir were due to encroachments of River

Banks reduced the water flow channel and thus washed away the constructions and thus property and people. In all these cases, even after weather forecasts warnings disasters have been taken place.

East Coast of India is faced by floods fury from the cyclonic activity. Mangroves used to act as protective walls of this tidal fury. With the passing of time, the mangroves were cut for building aquaculture ponds, construction activities by oil & gas companies, and others in violation of Coastal Zone regulations. Hotels, tourism, sand removal, and other encroachments of CRZ areas helped breaching of the coastal zones. All these created a pseudo sea level change by inundating coastal low lying areas with salt water.

Silt depositing at the mouth of the river joining the Seas and population establishing settlements around these areas for their livelihood were severely affected by the cyclonic tidal waves/flood waters. Here, let me give a classical example of Bhopal Gas Tragedy. When the factory was established, people flocked to the industry boundary wall to start small shops to meet their earnings. Though there is a prohibition to establish hutments within a kilometer from the boundary, the government officials failed to implement this law and that lead to great human tragedy. This is not an isolated case, but it is a fact with most of the factories.

If we look at UNICEF Report "The impact of climate change on Children", Nov. 2015, unfortunately, the UN organizations, World Bank, other funding agencies for third world nations are desperate to show they are in line with UN agenda by wasting public money for bringing out such poor quality reports. They attributed the issue pertaining to poor governance by individual nations to climate change, a de-facto global warming, wherein global warming is insignificant to influence children [rich or poor]. They state that "There may be no greater, growing threat facing the world's children and their children than climate change" though the new survey in US considered climate change a threat by only 3% of the surveyed people. Household air pollution is not climate change, floods and droughts are part of natural rhythm in climate this was there in the past, will be there in future. Extreme temperatures are region specific they are associated with the general circulation system of a region.

Instead of harping on carbon dioxide, let us work on controlling pollution [air, water, soil & food] that is affecting humans and other life forms on the Earth, which may aggravate if not controlled. Groups use beautiful jargons of words and at the same time rhetoric sentiment on emissions. Climate change is not affecting global water cycle but global water cycle is a part and parcel of climate change and climate change is a part of climate system. They vary with space and time, which are beyond human control. However, limiting the impact is in the human hands.

Sustainable development goals stable biosphere at risk: they relate to pollution [air, water, land and food] rather than carbon dioxide that provides carbon credits share from green funds of billions/trillions of dollars. Humans including all life forms of biosphere are suffering and severity of sufferings will increase with the time further under unabated growth of pollution. To achieve sustainable development and to protect the biosphere from risk, we must ask our governments to bring pressure on UN to include pollution and not carbon dioxide in their agenda. Unless we do this, we find in coming decades more crippled population and more hospitals, more Pharma companies to pollute the environment more and more.

Good governance plays an important role in this aspect of vital importance to individual nations. To achieve this goal, individual national governments must differentiate between the vested interests from peoples' needs. As long as national governments serve vested interests like that seen in India, people have to suffer. The sufferings will increase with the time.

## 5.5 Impacts on Nature

### 5.5.1 Is Global Warming a Reality?

1. The use of average of a wider area to study its impact on localized areas like agriculture is bad in science as the average present wide space and time variants in it. Global average temperature anomaly consist of wide variants in terms of climate system and global circulation patterns wherein wind speed and direction plays an important role. Such an average has no significance when we assess the impact on changes in weather, ice and in sea level and as well on water resources & agriculture, which are location-

region specific. Though we are talking of anomaly, they present high variations in space and time: ocean to land; Southern Hemisphere to Northern Hemisphere; country to country and region to region within a country; location to location within a region.

Global warming is put on false foundations. In the literature, it is commonly seen that the word “climate change” is misused as de-facto “global warming”; and “the global [land & ocean] average temperature anomaly” is misused as de-facto global warming. Global warming refers to temperature while climate change refers to meteorological parameters, more specifically to precipitation and temperature. Precipitation presents fluctuations but they are region specific, following general circulation pattern-climate system. This modifies the temperature accordingly. Also, unusual cold & warm conditions prevailing over different parts of the globe are associated with local general circulation patterns and associated multi-decadal Oscillations/Southern Oscillation. For example, Polar Regions are affected by the frequency & duration of occurrence of circumpolar Vortex [6 months day and six months night] and in India the location of High pressure belt around Nagpur in summer & winter. They are primarily related to wind.

IPCC in its AR5 stated that more than half of the global [land & ocean] average temperature anomaly is associated with the “greenhouse effect” component [global warming is a part of it] and the remaining [less than half] component is associated with the “non-greenhouse effect” [ecological changes]. It must be noted the fact that: the urban-heat-island effect with its’ dense met network overemphasized and rural-cold-island-effect with less dense met network is underemphasized in the global [land & ocean] temperature anomaly and thus as a result it is over emphasized; but this is not so with the satellite data. The model based predicted temperature anomaly curves are compared with global average temperature anomaly but not with global warming component. Based on that present predictions on expected temperature in 2100 or doubling of carbon dioxide in the atmosphere with upper and lower limits around the mean. They are hypothetical far from reality and using such predictive range, postulated the sensational impacts on nature.

The consensus [the average] of the climate models used by the IPCC for their 5$^{th}$ Assessment Report shows no long-term increase in global precipitation from 1861 [the start of the mean of the climate model outputs] to 1999 [the end of the 20$^{th}$ century] though their global average temperature anomaly during the same period presented an increasing trend. However, increase or decrease in precipitation shows a decrease or increase in temperature. For example, during the 2002 and 2009 drought years in India, the annual average temperature has gone up by 0.7 and 0.9 °C.

Sea Breeze and land Breeze patterns over East and West Coasts of India are different. Here we can get the clue from the human comfort equation wherein under the same temperature & humidity conditions with changing wind speed [direction also play an important role based on the location] the human comfort changes. Same way, heat and cold waves, wherein the high pressure belt location and winds decide the direction in which they move and create the heat wave condition in summer and cold wave condition in winter. It is like advection of heat or cold based on winds direction & strength.

Though the precipitation presents fluctuations with different periods, in a rare case of solidarity they all present below or above the average precipitation condition. For example, in the Southern Hemisphere, the fluctuations, though, cycles varied from 52 to 66 years [the main cycle], they all coincidentally come under the below the average condition in 2013. This resulted above the average temperature anomaly in Southern Hemisphere. However, this will be part of natural variation superposed on trend.

The central England the longest continuous instrumental record of annual mean temperature shows 0.95 °C rise during 356 years [1659-2016]. The peak similar to that observed after 1990 was also presented after 1730.

NOAA satellite data of sea ice extent from 1972–1990 presented a fluctuation pattern but Northern Hemisphere fluctuation pattern is in opposite direction to that of the Southern Hemisphere fluctuation pattern. That means in one hemisphere when ice is diminishing in the other hemisphere ice is increasing.

2. There is a big discussion on: Why there are differences between surface temperature measured in Stevenson Screens and atmosphere as measured by satellite. Here we are not talking of station value but a region value and the averaging over that region. Also, we are talking of trend and

not the fluctuation part. To get meaningful answer to such questions we must compare and explain for the causes of differences.

Christopher Monckton of Branchley [7 January 2016] observed that "When the first temperature monitoring satellites began to deliver data, NASA said the satellites temperature records would be more reliable than the surface record because the coverage was more complete, the method of measurements standardized and the coverage and coverage-bias uncertainties that plague the terrestrial record were absent. Schmidt & Karl say there has been rapid warming around the past 19 years. If so, where on Earth did it come from? The official theory is that $CO_2$ warms the atmosphere and the atmosphere warms the surface. But for almost 19 years the satellites show that the lower atmosphere has barely warmed. Some argued that the earth is influenced by the Sun. Here all greenhouse gases normally influencing earth-atmospheric temperature were not clubbed with the $CO_2$ as we are discussing the anthropogenic greenhouse gases affect global warming.

IPCC had predicted in 2007, on the basis of paper by Ben Santer that the tropical mid-troposphere should warm twice or even thrice as fast as the tropical surface. However, according to Schmidt & Carl the predicted tropical mid-troposphere hot spot is in reality absent. Lower and mid-troposphere anomalies are almost identical. Aside from the ocean warming, the land based warming was prominent over Siberia and northern Chine, Europe and Central America, owing much too urban heat-island effects. The ARGO bathythermographs show little warming at the ocean surface from 2004 until current El Nino began though generally greater with the depth and the question is how greenhouse warms the deeper layers leaving the surface?"

After the El Nino in 1997/98 the global average temperature anomaly has gone up and maintained that trend under the ground based data series; but under satellite data series the temperature anomaly has come down and presented no trend. Generally, on either side of El Nino peak show depressions. Under the Southern Oscillation, the temperature rises during El Nino phase and thereafter the temperature drops down and later recovers to normal condition. However, the changes in temperature associated with the Southern Oscillation and volcanic activity becomes part of intra-annual and intra-seasonal variations and thus their contribution to long term trend is insignificant. By fitting such data to sine curve or moving average, we get the realistic trend as explained earlier.

Reports state that "RSS satellite monthly global mean surface temperature anomaly data set shows no global warming for 18 y 8 m since May 1997, though one-third of all the anthropogenic forcing have occurred during the period of pause. The UAH satellite data showed a pause almost as long as the RSS data set. However, the much altered surface datasets show a warming rate 1.1 °C per Century during the period of pause for May 1997 to September 2015 after the data adjustments. During 1997-2016 the difference in the warming trend between RSS [satellite data] and GISS [surface data] is around 0.25 °C. Also, Global Temperature trends [Phil Jones] repeated with around 0.15-0.16 °C during 1860-1880, 1910-1940 & 1975-2009. Between 1979 and 2014 two volcanoes have occurred, namely Chichon in 1983 and Pinatubo in 1991 presented cooling effect prior to 1997/98 El Nino. However, the data from 1979 show a Sine Curve pattern with zero "trend" in both the surface data and satellite data. As the data sets for truncated part of the cyclic pattern, they both show an increasing trend [lower peak to upper peak in a sine curve], though it is zero trend for the complete cycle.

3. When the data series show different patterns, blindly fitting the data series to a linear curve it gives misleading conclusions. For example, if we use the data of 1978 to 2015 [444 months or 37 years], UAH satellite data presents a trend of about 1.14 °C per Century. After seeing this trend that contains period before and after 1997/98 El Nino, I pointed out saying that while fitting such data, we must eliminate the anomaly associated with El Nino affect or fit the data sets prior and after El Nino periods separately to see the trend. I have seen this in https://kenskingdom.wordpress.com/ an article "Earth and Water" on January 13, 2016. The analysis of both the periods namely prior and after 1997 showed a "nearly zero trend". Some mentioned that prior to 1997 volcanic activity cooled the temperature and after 1997 the raised temperature did not come down in surface data but in satellite data it did come down and followed a zero trend pattern. If the difference is zero at 1997, it is 0.25 °C at

2015. That is the difference in around 18 years is 0.25 °C and thus the surface data showed an increase of 1.1 °C per Century. In the North Extra Tropics: the differences between the prior and after 1997 data series of 229 and 216 months data series with zero trend showed a sudden jump around 1997 by 0.48 °C on land 0.26 °C on ocean. This is the difference between volcanic cooling and El Nino warming on land and in ocean. **Table 5.1** presents the differences under three surface measurements and two satellite measurements in three different periods:

From the above table it is clear that in all the data sets, they presented a decreasing trend per Century. Can we call this a non-linearly decreasing trend in the contribution of anthropogenic greenhouse gases effect?

**Table 5.1** Variation of global average temperature anomalies with type of measurements & period of data

| **Measurements** | **1990-2015** | **1995-2015** | **2001-2015** |
|---|---|---|---|
| | **[°C per Century]** | | |
| **Surface measurements** | | | |
| GISS | 1.71 | 1.51 | 0.76 |
| HadCRUT4 | 1.62 | 1.52 | 1.11 |
| NCEI | 1.58 | 1.49 | 1.40 |
| **Average** | **1.64** | **1.51** | **1.09** |
| **Satellite measurements** | | | |
| RSS | 1.11 | 0.42 | - 0.11 |
| UAH | 1.00 | 0.38 | 0.12 |
| **Average** | **1.05** | **0.40** | **0.00** |

Some argued that the atmospheric temperature anomalies are necessarily different from surface anomalies. Usually, atmospheric anomalies are less than the surface maximum in hot periods and higher than the surface anomalies in cooler periods. It is like night and day conditions. We need to average them and thus they should present the same averages both surface & satellite measurements.

Here we must remember the fact that the surface data does not cover the entire climate system and general circulation patterns. Particularly in land areas, the met stations are dense in urban areas and sparse in rural areas this is clearly evident in Australia's met net work and also the periods are different. Satellite data covers all the climate systems and general circulation patterns. Thus, surface data is more biased with urban conditions and thus expected high global temperatures. This is what the present scenario. Lowering the past data series and raising the current data series also add to this. Naturally surface data will be higher than satellite data.

From all these, it is clear that, how you analyze the data series plays the critical role in getting correct solution. Blind application of linear curve fitting leads misleading conclusions when the data showed cyclic pattern.

4. What we need to address now is: study the local & regional temperature patterns and look in to causes for regional and local differences. Then only it serves the needs of policy makers to common man. Without that, simply harping on global average will lead to disasters in coming decades with trillions of dollars spending on good for nothing theory of "Global Warming & Carbon Credits".

On this observation of mine in "Watts Is Up With That", John Knight (January 14, 2016) observed "Thank you for trying to get the science back on track, and back to work trying to help real people with real need of useful climatology/meteorological information and research. No doubt some in the field have never stopped in the effort you champion here, and it's a shame they are being drowned out by this geotesque caricature of real science they call climate change"

**Reality:** The localized greenhouse gases [natural and anthropogenic] interact with the localized energy component under location conditions, resulting in change in temperature. To this temperature change added several other components, namely:

- Advective energy, associated with general circulation patterns,
- Associated with the climate system,
- Non-greenhouse component, associated with the urban-heat-island effect, rural-cold-island effect, filth cover effect, deforestation or reforestation, road network, etc,

- Natural variability component [short & long term], associated with Southern Oscillation, PDO, AMO, and other localized oscillations, droughts & floods in precipitation cycles, etc,
- Irregular variations component, associated with the Southern Oscillation, volcanic eruptions, wars, etc.

5. Warmists, Warmist Institutions, Warmists Media are desperate to show (+1) + (-1) is equal to (+2) and don't like to accept it is (0). The same we can see with reference to 2016 "Nor-easter" that drenched northeastern US with blizzard. They are normal for that region and a separate entry is presented in Wikipedia on it. NCDC/NOAA reports that blizzard of 2016 makes 4$^{th}$ among the worst winter storms of past 100 years. Historical data records show blizzard of 1888 followed the path of 2016 blizzard; but it is a much greater storm than 2016 blizzard with lower temperature regime. Temperature is not the real driving force. For example India receives severe cyclonic storms in pre-monsoon, monsoon and post-monsoon seasons with declining temperature. They are highly associated with general circulation patterns that provide the wind and temperature gradients. Warmist media poured volumes on 2016 blizzard attributing it to global warming, a non-entity. The same is the case in India wherein the PR groups of IPCC/Al Gore rush to the media with the stories attributing the disaster event [whether it is a drought or flood or heat wave or cold wave] to global warming. I write letters to government agencies including the Prime Minister of India clarifying such misgivings and the same I sent to media but nobody likes to present that.

The average glacier has lost about 39 feet or 12 meters, of ice thickness since 1997, according to Samuel Nussbaumer at the world glacier monitoring service in Switzerland but that organization has taken measurements at only 230, or one-seventh of one percent, of the world's 160,000+ glaciers. It has done very little work in Antarctica, which contains around 90% of the world's ice mass, and which, except for a small area of West Antarctica, has not warmed or lost ice mass throughout the satellite era.

Satellites [mean of RSS and UAHv6] show no global warming since April 1997. So, none of the changes that have occurred could have been caused by warming. Also, IPCC, both in its 2012

Special Report on Extreme Weather and in its 2013 Fifth Assessment Report, says there has been no particular overall trend in storminess, floods or droughts. In the case of sea level rise the observational errors are far higher than the mean rise.

6. **Here it is pertinent to present Dr. Patrick Moore founder chairman of Greenpeace organization observations from a presentation of 2015 Annual GWPF Lecture titled "Should We Celebrate Carbon Dioxide?"** at the Institute of Mechanical Engineers, London 14 October 2015. He argued that: "As I have stated publicly on many occasions, there is no definitive scientific proof, through real-world observation, that carbon dioxide is responsible for any of the slight warming of the global climate that has occurred during the past 300 years, since the peak of the Little Ice Age. If there were such a proof through testing and replication it would have been written down for all to see. The contention that human emissions are now the dominant influence on climate is simply a hypothesis, rather than a universally accepted scientific theory. It is therefore correct, indeed verging on compulsory in the scientific tradition, to be skeptical of those who express certainty that the science is settled and the debate is over. But there is certainty beyond any doubt that $CO_2$ is the building block for all life on Earth and that without its presence in the global atmosphere at a sufficient concentration this would be a dead planet. Yet today our children and our publics are taught that $CO_2$ is a toxic pollutant that will destroy life and bring civilization to its knees. There was no guarantee that fungi or any other decomposer species would develop the complex of enzymes required to digest lignin. If they had not, $CO_2$, which had already been drawn down for the first time in Earth's history to levels similar to today's, would have continued to decline as trees continued to grow and die. That is until $CO_2$ approached the threshold of 150 ppm below which plants begin first to starve then stop growing altogether, and then die. Not just woody plants but all plants. This would bring about the extinction of most, if not all, terrestrial species, as animals, insects, and other invertebrates starved for lack of food. And that would be that. The human species would never have existed. This was only the first time that there was a distinct possibility that life would come close to

extinguishing itself, due to a shortage of $CO_2$, which is essential for life on Earth.

World Bank brought out a report on Decarbonizing Development. The whole report's validity depends up on what they mean by 2 °C rise. Is it due to global warming or trend associated with several other factors represented by global average temperature anomaly in which local and regional and natural components or forcings are involved? They did not touch up on this point. That makes the report good for nothing report. The second issue is by keeping aside the high power consuming IT sector from the purview of reducing carbon also has no meaning. World Bank is shy of using the word global warming, though it talks of carbon dioxide but uses climate change, which has a broader meaning. There is a need to educate all such international bodies on the use of such words.

7. The word "global warming" is a misnomer. This is like "digging a mighty mountain and catching a rat". There is nothing like "global" warming, as they all refer to synthesized local/regional warming-cooling components only. This has no affect on nature at global scale. Such a synthesized local/regional temperature will not affect the nature but the local-regional weather affects nature around the local-regional zones.

Global average is an unethical average. For example, it is like average income of a global family or man or women from the data: 54% of the global wealth is in the hands of 64 individuals and the rest of 46% of the wealth in the hands of the rest of the people [99.999%]. In global average for the month of January is the average of January in the NH [a winter month] June of SH [summer month]. More area in the SH is covered by Oceans and more area in the NH is covered by land. The average will have different meanings and different impacts on nature in those two Hemisphere conditions. Same is the case with region to region or location to location.

### 5.5.2 Are the Impacts of Climate Change on Nature a Reality?

1. Under weather & climate, everything has become unusual and try to attribute to something we don't have the comprehensive knowledge like "a blind man using a light

pole". We never care to look in to history. Most unfortunate thing is that the man on the street to the Editor of scientific Journals/magazines talk of climate change at the same wavelength, and thus science of climate change became a scapegoat. Thus, the "Science of Climate Change" has turned in to political satire of "Global warming & Carbon Credits". IPCC is giving weight to number rather than to science. In this process the major casualty is the health of life forms on the Earth, more particularly in developing countries, most of which are located in "Warm Tropics". All these are happening even after IPCC changed its tone and tenor from AR4 to AR5 on global warming and its impact issues after 2009 December Copenhagen fiasco.

The word climate change is widely misused. Majority are using it as an adjective to get hype to their reports or to get published in reputed scientific journals & in media and through which get funds; and majority of the scientific community use climate change as de-facto global warming as they are shy of using the term global warming as they were not confident on its % share in global [land & Ocean] temperature rise quantitatively. In this area with billions of dollars are available for research and carbon credit policy. Reputed national and international journals and media are publishing articles and reports with highly misleading headings. Through these media they are attacking the opposition to the pseudo global warming and its impact on nature.

However, it is cautioned here that heuristic estimates based on models may lead nowhere but provide a means of sensationalize inaccurate-misleading conclusions. We must realize the fact that "Ignorance is terrible but exaggeration is dangerous". Most of the inferences on impacts made by IPCC were based on such hypothetical models only to create an atmosphere of sensationalization fear among rulers and people; and not based on ground facts relating to general circulation patterns existing over different climate systems of the globe.

After December 2009 Copenhagen Climate gate, UN Secretary General Ban Ki-Moon said "IPCC should have stronger scientific basis for making it predictions and recommended an overhaul of the position of Pachauri, the Chairman of IPCC.

2. IPCC's Fifth Assessment Report (AR5) consists of three Working Group (WG) Reports and a Synthesis Report. The

first Working Group Report was published in 2013 and the rest were completed in 2014. They are:

- WG-I: The Physical Science Basis released on 30 September 2013, Summary for Policymakers published on 27 September 2013
- WG-II: Impacts, Adaptation and Vulnerability released on 31 March 2014
- WG-III: Mitigation of Climate Change released on 11 April 2014
- AR5 Synthesis Report (SYR) released on 2 November 2014

It is quite conspicuous to see nowhere IPCC referred the word "global warming" and in its place invariably used climate system, climate change with several adjectives attached to it.

Is IPCC using climate change synonymous to global warming? The answer is yes. The so called impacts on nature relate to global warming or climate change? The answer is climate change. Global warming is a trend in temperature; and climate change is a rhythmic variation superposed on trend, in which extremes form part, in temperature and rainfall. During drought years the temperature and evaporation goes up and during flood years the temperature and evaporation goes down. Studies in early 80s showed that cube root of precipitation follows the global solar radiation and evaporation. Because of this interrelation between rainfall and temperature, the rhythmic variation present opposite patterns. Studies in late 80s showed that Indian southwest monsoon precipitation and thus the global temperature followed a 60-year cycle.

After receiving Noble Prize, both IPCC and Al Gore withdrew their conclusions on Himalayan Glaciers melt and Arctic ice. IPCC admits glacier goof-up and now says the Himalayan Glaciers won't melt by 2035 and observed that "It has come to our attention that refers to poorly substantiated estimates of rate of recession and date for the disappearance of Himalayan Glaciers. In drafting the paragraph in question, the clear and well established standards of evidence, required by IPCC procedures were not applied properly". Al Gore admitted similarly mistake of his own observation that "the latest research showed that the 'Arctic' could be completely ice-free in five years" is not based on science but based on "assumed" information like that of IPCC's.

IPCC's WG-II AR5 the second of the three assessment reports that assesses the impacts of climate change, adaptation and vulnerability of human and natural systems, runs around as highlighted by UN Climate Change secretariat in Bonn "The IPCC report makes clear that people around the world are already suffering from climate change, as it directly affects their livelihoods, reducing crops, destroying homes and rising food prices, and that this will accelerate if climate change is left unchecked it provides a detailed assement of regional aspects, which give a much more clear understanding of climate impacts in different regions". Touching the human conflict aspects, the secretariat said, "Among other things, the report warns that climate change increases the risk of armed conflict around the world because it worsens poverty and economic shocks. Therefore climate change is already becoming a turmining factor in the national security policies of states". McCathy, EPA/USA Administrator observed that "Carbon Dioide rise causes more extreme heat, cold, storms, fires and floods". All these refer to global warming only and climate change was used as defacto global warming. We have seen in the above presentations, the global warming is insignificant to influence nature as stated. Climate change includes other factors such as natural variability. They were there in the past and will be there in future. There are other localized components of climate change that influence these natural variations to a certain extent. Also as discussed above, the direct actions on nature are agreevating the disasters. For example, the desaster caused by 2015 floods during the Northeast Monsoon Season over Nellore in Andhra Pradesh and Chennai in Tamil Nadu in the Southeast Coast of India is a classical example. Even though such precipitation [amount & intensity] was not new, the destruction and people's sufferings were unprocedent because the rivers & drains that carry rain water and seewage and tanks that store rain water & domestic seewage were converted in to concrete jugle and thus water flooded the residential areas. This is nothing to do with global warming and yet several groups media tried to attribute the blame to global warming & climate change. Here the precautionary principle, in fact, should have saved the people. Here human greed overplayed the precautionary principle. This is the same in all urban areas in India. The poor town planning is causing the devastation. It is unethical to attribute impacts associated with poor town planning to climate change or global warming.

3. The Sun is the main driving force for weather on the Earth. Among the many components, sunspots present a clear cut cyclic variations running in to 10.0 ± 0.5 years and its multiples. Because of this, they present changes in amplitude from cycle to cycle. Similar pattern was observed in ground based total solar radiation [global solar radiation] and net radiation [radiation balance]. They are part of general circulation patterns.

The natural variability in principal met parameters, precipitation and temperature, follow the general circulation patterns existing over different parts of the globe, in climate system. The weather at Polar Regions are controlled principally by the formation of Circumpolar Vortexes, which is part of General Circulation pattern in which polar natural oscillations form part. Thus weather associated with these systems control the changes in ice over the North and the South Poles and thus associated Sea Level Changes if any. However, these are modified by human actions such as massive drilling for oil, gas & water, destruction of coastal zones, etc.

Historically floods and droughts affected civilizations. The same will be the case in future. By understanding the natural cyclic patterns in precipitation and snowfall, life must be adapted to such changes. Any activity that does not account these factors have to encounter disastrous consequences in terms water resources, agriculture, impact of disaster, etc.

4. Centre for Science and Environment [CSE] a New Delhi based NGO in its' on line Edition of Down To Earth magazine presented a story released on 15th January 2016 by FAO Representative in Ethiopia Anadou Allahoury. The Title of the article is "El Nino: Ethiopia faces worst drought in thirty years". The crop production in Ethiopia dropped by 50 to 90% in some regions and failed completely in the eastern part of the country. The outlook for 2016 is very grim. According to FAO Resident Representative "After two consecutive seasons of failed crops, the success of the main cropping season that starts now will be critical to preventing conditions from worsening". The prevailing El Nino phenomenon is caused crop failures, reduced livestock and pushed around 10.2 million people in Ethiopia in to food insecurity.

Unfortunately, it has become a ritual, to attribute every weather event to El Nino or Global Warming without looking in to weather & climate data of the regions.

WMO released a press note on extreme weather events of 2013 on the occasion of World Meteorological Day (23-3-2014) in which it observed droughts and warming conditions in Southern Hemisphere Nations. My studies of early 1980s on the rhythmic variations in precipitation data series of stations in Southern Hemisphere Nations and projections showed below the average precipitation during 2013. Though in the Southern Hemisphere nations presented cycles of different periods in the precipitation data series [52 to 66 years], coincidentally majority of them showed below the average pattern during 2013. The higher temperature regime in Southern Hemisphere Nations could be attributed to the dry weather prevailing in 2013. This is not part of global warming but it is a natural variations part of climate change. These influence the water resources availability and thus agriculture.

The Ethiopia's case, I as a Team Leader/Chief Technical Advisor of UNDP/WMO project [ETH/86/021-WMO/UNDP] in Addis Ababa/Ethiopia brought a report "Climate fluctuations in the precipitation data of Ethiopia during the period of meteorological record", Agroclimatology Series-4, on 1 June 1990, Addis Ababa, Ethiopia. The results of this study on climatic fluctuations in the precipitation data of Ethiopia suggests that at present in major part of the country is under below the average precipitation pattern. The predominant cycles are 22, 28, 36, etc. For example

- In areas with 28 year cycle [Jijiga, Wonji, Debre-Zeit, Bahir Dar low rainfall with bio-model rainfall with dry crops like Sorghum/Pearl Millets], new cycle starts around 1991. That means, around 2006 the below the average 14 year period started.
- In the case of areas with 36 years cycle [Gore, Debre Markes, Gambella, Jimma, Welkite, Combolcha, Dire-Dawa better rainfall with Coffee, Barley, Wheat] the new cycle starts around 1989. That means, around 2008 the below the average 18 year period started.
- In areas with 22 year cycle [Asmara now in Eritrea low rainfall], the new cycle starts around 1997. That means, around 2009 the below the average part started.

- Addis Ababa and Massawa [moderate rainfall with the staple food crop Tef] presented irregular variations being bordered to differing cycles with Nagella the below average 18 year period starts around 2019.
- Gonder presents a 52 year cycle with below the average part commencing in 2015 and will continue for 26 years with 6 years better rainfall in the middle: 10 year 6 years 10 years.

These years are given in my publication in a table [figures are presented for individual stations]. So, they are part of natural cycle and not due to El Nino. I visited all these regions.

Heatwaves and cold waves are part of General Circulation patterns. Same is the case with changes in ice and sea levels. Before we make sweeping conclusions on the impacts of the so called global warming and climate change, we must understand the climate system and general circulation scenario of the region.

Hundred of islands near the Equator in Papua, Indonesia, have been undercut by the sea in a manner that gives credence to the hypothesis that there has been little net change in sea level in the past thousands of years. It takes a long time for so much erosion to occur from gentle wave action in a tropical sea.

## 5.6 Summary and Conclusions

### 5.6.1 Summary-Synthesis

Global warming is a misnomer. Climate change relates to changes in meteorological parameters. Global averages have no impact at local-regional scale. IPCC in its AR3 definition of climate change used "changes in the composition of the *atmosphere* or in *land use*"; but to create confusion, under UNFCC Article 1, "atmosphere" is replaced by "*global atmosphere*" and instead of "land use", used "*directly or indirectly human activity*". Thus, as per AR3 definition, climate change is a local-region specific activity only; and thus its impacts on nature are felt at local-regional scale. Because of this the changes in sea level, ice melt, glaciers retreat, droughts-floods, cold-heat waves, agriculture-water resources, etc are all location-region specific activities only.

The film "An Inconvenient Truth" fetched Al Gore tons and tons of currency but failed the prophecies of the film "Dooms Day". Sundance Film Festival in January 2006 at a premiere show of his film declared "unless drastic measures to reduce greenhouse gases are taken within the next 10 years, the world will reach a point of no return". But, the Earth and humanity has been doing just fine since then except that, as usual, some areas are affected by floods and some others by droughts; some areas were affected by warm weather and some other areas were affected by cold weather, etc. A Case was filed against Al Gore's Sci-fi comedy horror movie "An Inconvenient Truth" in the London High Court in 2007. The judgment was in favour of the petitioner.

With the Copenhagen fiasco of December 2009 [I am one of the signatory a memorandum that was submitted to UN Secretary General]; IPCC changed its tone and tenor from 2007 AR4 to 2013/14 AR5. IPCC withdrew its conclusion given in AR4 on Himalayan Glaciers that "Himalayan Glaciers will melt by 2035" with regret. After returning from COP21 Paris summit on climate change, the Indian Minister of Environment, Forests & Climate Change informed to the Indian Parliament [Rajya Sabha] that, 86.6% of Himalayan Glaciers are stable. Al Gore withdrew his conclusion that "Greenland will become ice free in 5 years". All these retractions by IPCC and Al Gore took place only after receiving the Noble Prize.

The annual seasonal march of Greenland ice shows that they are within the double standard deviation around the mean. The summer low has gone down in 2012 and since then it is recovering and thus moving upwards. Even with the claims of 2015 as the warmest year, Greenland ice is moved up with the seasonal pattern intact.

In fact, IPCC-AR5 observed that "So far there are no widespread observations of changes in flood magnitude and frequency due to anthropogenic climate changes that is due to 'global warming'. Even their projections imply limited evidence. The projections showed flood hazard increases in "some parts" of the globe. This also presented limited evidence. Recent socio-economic losses from flooding have increased mainly due to "greater exposure and vulnerability". The drought conditions in Southern Hemisphere and associated warming; and 2014-2016 droughts in Ethiopia are associated with natural cycle in precipitation data series. These were predicted by the author of the book in 1980s based on his study on cycles in precipitation data series.

Wikipedia talks that nearly all scientists agree that "humans" are contributing to the observed "climate change", 97.2% supported the consensus view that it is manmade. Here it is pertinent to mention President Vladimir Putin believes that "there is no global warming, that this is a fraud to restrain the industrial development of several countries including Russia. We found that, while climate change does exist, it is cyclical, and the anthropogenic role is very limited"

The global average temperature anomaly consists of several factors wherein majority of them are location-region specific. Global warming component though used as global factor but in reality it is also a location region specific. Because of this, IPCC in its AR5 report made some qualitative statement on the % contribution of global warming component to global average temperature anomaly by saying that "the contribution of greenhouse effect component is more than half and the non-greenhouse component is less than half of the global average temperature anomaly. Since 1951, global warming, a part of greenhouse effect, contribution started. The other part of greenhouse effect relates to aerosols which were there in the atmosphere like natural greenhouse gases but recent years [after industrialization] several other man induced aerosol components are added to natural aerosols. Over the oceans and on land filth plays an important role in controlling energy absorption and reflection.

Here there is one fallacy: more than half is shown as global warming by showing other anthropogenic forcings as around zero. Here, the other anthropogenic forcings is shown as due to aerosols from volcanic eruptions. Aerosols not only act in cooling but also it reduces the energy reaching the Earth and thus global warming component.

Reports suggest that average global ocean temperatures are lower than land surface temperatures. They are around 0.4 °C and 1.7 °C, respectively over oceans and land that is land + ocean is 0.79 °C. Since 1979, the land temperature has gone up by about 0.25 °C and ocean temperature by about 0.13 °C that is land + ocean temperature has gone up by 0.116 °C [0.46 °C per century]. Here we must remember a fact that when we fit a truncated data series that follow cyclic variation to linear regression the trend will give different results based on which part of the cyclic variation the truncated data forms [**Figure 4.4**]. Here the results present the increasing segment of the cyclic pattern data series. Prior to 1997/98 presented zero-trend and after 1997/98 also presented zero-trend. That means the global average temperature anomaly presented sudden jump by 0.46 °C per century instead of zero. This is the fallacy of blind fitting the data to linear trend.

Let us see **Figure 1.6a**, it presents the global yearly mean surface [land + ocean] temperature anomaly. The data was fitted to linear trend and multi-decadal oscillation [60-year cycle]. The multi-decadal oscillation of a 'sine curve' varied between 0.3 and +0.3 °C. The trend showed an increase of 0.0059 °C per year starting from 1880. Thus from the base year 1880, by 2015 the anomaly trend rises to 0.8 °C [above it is given as 0.79 °C] and by 2100 it rises to 1.3 °C. IPCC considered 1951 as the starting year for global warming. That means, the rise between 1951 to 2015 and 1951 to 2100, respectively are: 0.4 & 0.9 °C during 2002 & 2009 drought years the all India mean temperatures have gone up by 0.7 & 0.9 °C. These are global warming period component. However, the satellite data shows these as one-third of them. However, the anomaly during the global warming period is influenced by several localized factors, namely:

- If we look at model predictions of global temperature anomalies, they presented a change in the pattern wherein with the passing of time the change in temperature anomaly with increasing carbon dioxide is tapering decreasing per unit of carbon dioxide with the time. That means anthropogenic greenhouse gases increase will have null effect on energy conversion in to temperature rise.
- Prior to 1960's on land, data was available at sparse network of unevenly distributed met stations in space and time. After 1960's though the number increased but the instruments changed in majority of them; and locations do not fit in to the prior to 1960's system.
- The met network on land was more concentrated in urban areas and sparsely over rural areas. These two are related to heat-island and cold-island effects. Thus, with this network heat-island effect is overemphasized and cold-island effect is underemphasized resulting, over emphasis of ecological changes [non-greenhouse effect] component in the global averaging of the temperature.

- Both the land & oceans are covered by filth, also they will present increasing trend with the passing of time with greater human activity. This will change the radiation balance at the Earth and the Atmosphere Absorbs less and reflects more.
- Oceans cover around 70% of the globe. Prior to 1990's the data was collected by ships on their way very poor quality coverage. After 1990 buoys were established but they are uneven in terms of space and time.
- Ecological changes component over oceans relate to covering of water by filth.
- Aerosols of volcanic eruptions along with several other aerosols and filth cover change the temperature. This automatically reduces the energy available for greenhouse effect at ground level [land & oceans].
- Natural variability that includes irregular variations is superposed on the trend and thus it also forms the part of anomaly.

We don't have the quantitative estimate of all these factors, including the so called greenhouse gases or global warming as IPCC through trial and error goes on reducing sensitivity factor from one report to the other from AR4 to AR5 it was reduced from 1.95 to 1.55. Also, to fit such trail-and-error process the ground based temperature data anomaly prior to 1950 was lowered and after 1950 the temperature data anomaly was raised. Also, during pause period again the temperature was raised satellite data shows hiatus/pause.

Global warming is used to sideline the main and real issues of climate and its changes with the time. Global warming is an average of cocks and bulls. Over the globe the temperature presents wide variations with space and time. These are associated with local and regional players but not global players. Thus, global "warming" is a misnomer. Global warming is not warming at global scale but it is sum and total of what is happening at local and regional levels only. So, *global warming figure is a misrepresentation of what is really happening over different parts of the globe and thus their respective impacts on nature around them and not associated with changes on nature elsewhere.*

Global warming refers to change in temperature but climate change refers to all meteorological parameters, which are interrelated-interactive. Global warming refers to trend in temperature but climate change refers to both trend superposed on it the systematic variations in temperature as well in precipitation. Therefore, we must stop referring to climate change when we are dealing with subject related to global warming that is, we must stop using climate change as defacto global warming.

IPCC-AR5 observed "The effect of climate change on crop production is evident in several regions of the world with high confidence. All aspects of food security are potentially affected by climate change, including food access, utilization, and price stability with high confidence". Agriculture is a local-region specific activity and they depend upon the local weather. Thus they are part of climate change. As per as others [other than production], they are primarily relates to governance [good or bad] and has nothing to do with climate change.

IPCC-AR5 observed "Warming trends and increasing temperature extremes have been observed across most of the Asian regions over the past century with high confidence". These are local-region specific activity and thus they are part of climate change. For example, in 2007-08, Andhra Pradesh state in India received bumper paddy production; but with the illegal export of rice caused high rise in price of rice.

The impacts of climate change on nature were there in the past and will be there in the future. However, they will be affected by human actions either through their greedy actions and or indirect actions through changing land use and land cover, water use and water cover and other physical actions.

Media must discourage presenting stories on individual weather events as manifestation of global warming. Weather events are part of systematic variations present in the climate change.

India with 17.5% of world population, whether rich or poor, are suffering from pollution [air, water, soil & food] but not from carbon dioxide and yet United Nations is more worried on Carbon Dioxide over pollution to protect the multinational companies that are destroying environment with their poor technological innovations, including IT [high power consuming] with severe health hazards.

Seasonal variations are far higher over year to year variations including unusual events in meteorological parameters as well as associated parameters such as Arctic and Antarctic ice changes, sea level changes, etc. Even with reference to El Nino we look at warming component and its impacts but rarely look at equally important parallel cooling zones and their impacts. Also, before and after the El Nino peak we notice drop in temperature.

Historically our forefathers adapted agriculture farming systems and water resources use to climate change. With the creation of IPCC the scenario changed. They have been polluting the science of climate change to protect multinational companies that were/are destroying the environment on the name of innovative technologies and minting trillions at the cost of human sufferings. Poor is becoming poorer and rich are becoming richer. Here it is pertinent what Mahatma Gandhi said “India has enough resources to fulfill everyone’s needs but it can’t fulfill one person’s greed”. This is quite valid for the world as a hole. Around 54% of the world’s wealth is in the hands of 64 people.

There is a need to characterize the factors associated with rise and fall of ice cover in each of the regions and this will facilitate to forecast the future scenarios. If we simply harping on global warming we never reach the reality; and thus their implications on sea level changes over different parts of the globe.

There exists several factors that influence Ice changes and sea level changes and they vary from location to location. The major issues in this direction are the earthquakes, volcanoes, human actions & extra-terrestrial actions relating to physical impacts, etc.

**Most Important Scenarios**

1. The warming theory in fact counters the acidification theory as it is clear from the historical temperature versus carbon dioxide pattern [**Figure 1.12**], wherein warmer water releases carbon dioxide in to the atmosphere and thus warming and acidification will not go together. The other important scenario is that the sea level rise, wherein sea level rise is caused by melting of polar ice masses. This way we are adding more water to the Oceans. Thus, this not only cool the Ocean waters but also dilutes the carbon dioxide concentration and thus counters the theory of acidification, though cooling allows more carbon dioxide absorption by oceans. The third scenario is the hiatus, wherein three theories were put forth and one such theory is that warming is hiding in the oceans. If the warming is hiding in the

oceans, it releases more carbon dioxide and thus more global warming and this must reflect on ground data. The fourth scenario is the natural multi-decadal and Southern Oscillation impacts. The fifth scenario is filth covering the oceans. Unless all these are integrated, the acidification & warming theories are of little significance.

2. **Figure 2.18** presents an actual way of averaging for a calendar year from the Southern Hemisphere and Northern Hemisphere monthly averages. The basic question we should ask the IPCC and other UN agencies is how this average influences the nature at any given location or region? Impacts on nature are season, region & hemisphere specific. When we average January month of NH with June month of SH and goes on to average of December month of NH with July month of SH, how they cause heat or cold waves; drought or floods; ice melt or sea level change; unusual events? All these are built on false foundations. Thus, climate change is local-region specific activity with no global implications.

Day by day temperature data is mutilated to serve the vested interests. We must concentrate our research using single station data [un-mutilated] to provide a guide to people. In this direction, agroclimatological an agrometeorological studies are of utmost importance. Let us pray UN comes in the right track and help nations in the right perspective, such as:

- To understand the reasons for change in sea level at any given area, characterize the expected causes associated with that in terms of weather [natural variability], human actions [destruction of coastal zones, drilling for oil, gas & water], natural disasters [cyclonic activity, earthquakes, volcanic activity], etc;
- To understand the reasons for changes in glaciers [thickness & Spread] at any area, characterize the expected causes associated with that in terms of weather [natural variability], human actions [sports, pilgrimage, developmental activities, etc], natural disasters [earthquakes, volcanoes, extra-terrestrial bodies], etc;
- To understand the reasons for changes in polar ice, characterize expected causes associated with that in terms

of weather [natural variability], human action, natural disasters, etc;

- To understand the reasons for changes in drought and floods at any given location and homogeneous zones, characterize the weather in terms of natural variability in precipitation and probable disruptions in that due to deforestation & ecological changes under climate system;
- To understand the reasons for changes in heat & cold waves at any given region, characterize the weather in terms of general circulation patterns of inside and outside that region;

## 5.6.2 Conclusions

1. **Data**: To prove the theories related to global warming and thus climate change and their impacts on nature needs an accurate, evenly distributed in space and time network of stations that measure temperature and greenhouse gases at global scale. They must cover climate system and as well general circulation system. There is no such network before 1960s and even after 60's, though, the network has been improved in number but they are not uniformly covered over space and time. Since 1979 though satellite measurements are available that covers climate system and general circulation system, warmists and modelers are not accepting them by saying that satellites do not measure temperature as they present far lower temperature trend over the ground based temperature trend. This must be resolved. However, individual nations must build accurate met data series for individual stations without indulging adjustments and provide the same for research institutions in the respective countries for agro-climatic and agro-meteorological analysis to assist in water resources and agricultural planning on which local economy depends.
2. **Global Warming**: From the definition of IPCC that the increased global average temperature anomaly since 1951 has two parts, namely (1) one caused by the greenhouse effect and (2) the other caused by non-greenhouse effect. The former's contribution to the global average temperature since 1951 is extremely likely that more than half; and thus the latter's contribution is less than half. Also, at the same time the anthropogenic greenhouse effect component has

two components, namely 1(a) the anthropogenic increase in greenhouse gas concentrations and 1(b) the other anthropogenic forcings. 1(b) is associated with volcanic eruptions and associated aerosols a natural phenomenon and man induced aerosols. (2) For this a new entry is filth covering oceans & land surfaces. However, IPCC is not sure of the quantitative contribution of these three groups on the global temperature. Thus, global average temperature anomaly is not global warming.

Historical temperatures versus carbon dioxide relationship to date show an opposite to Global warming theory, wherein warmer waters of the ocean release carbon dioxide in to the atmosphere and cooler waters of the oceans absorb carbon dioxide from the atmosphere with a time lag. Thus ocean acidity changes accordingly. Also recent studies showed the carbon dioxide life in the atmosphere is limited to 40 years and not to 1000 years. Also, IPCC through trial and error process continuously is reducing the sensitivity factor of converting energy in to temperature in carbon dioxide bands.

The global average temperature anomaly presented hiatus/pause for the last 19 years. *Though the global average temperature anomaly was built from spurious data, data managers to remove this hiatus-pause adjusted the data series. The data managers lowered the past data and the current data was moved upwards. This type of data manipulations also effects the natural inbuilt oscillations and thus affects the weather forecasting mechanism world over. This will be a major disaster in coming decades, unless we stop harping on global warming and its impacts. We must remember the fact that the global average temperature anomaly doesn't reflect the condition of a location-region and thus its impacts on nature at that location-region. To get better insight in to the problems and thus take remedial measures, the location-region specific studies must get priority.*

Thus there is no global warming as such. The word global warming is a misnomer. The global average temperature is sum and total of widely varying patterns over the globe. This average consists of several components with location-region specificities. Global warming component is only one part of this average and that too it is also a location-region specific activity with negligible contribution beyond 280 ppm as all the light emitted by the surface in the strongest carbon dioxide bands was completely absorbed at pre-industrial levels [< 280 ppm] and thus the Earth stabilized at 15 °C.

3. **Climate Change**: Climate change is real. Climate change is not global warming. Climate change was there in the past and will be there in the future. We need to adapt to them. Climate change is highly location-region specific, interwoven with the climate system and general circulation patterns. They form part of all meteorological parameters, more particularly precipitation and temperature. Wind & pressure are part of general circulation pattern. Indian rainfall is topography specific, which is part of climate system. Natural variability form the main part of climate change. The local-regional changes do influence these natural cycles.

"May 2016 floods in Seine River that runs directly through the heart of Paris were attributed to climate change. In fact 1910 floods were more devastating than of 2016 [**Figure 5.3 a & b**] – the level of water in the river were far higher in 1910 compared to 2016 with sedimentation of the riverbed. There were several other severe floods before 2016.

[a]

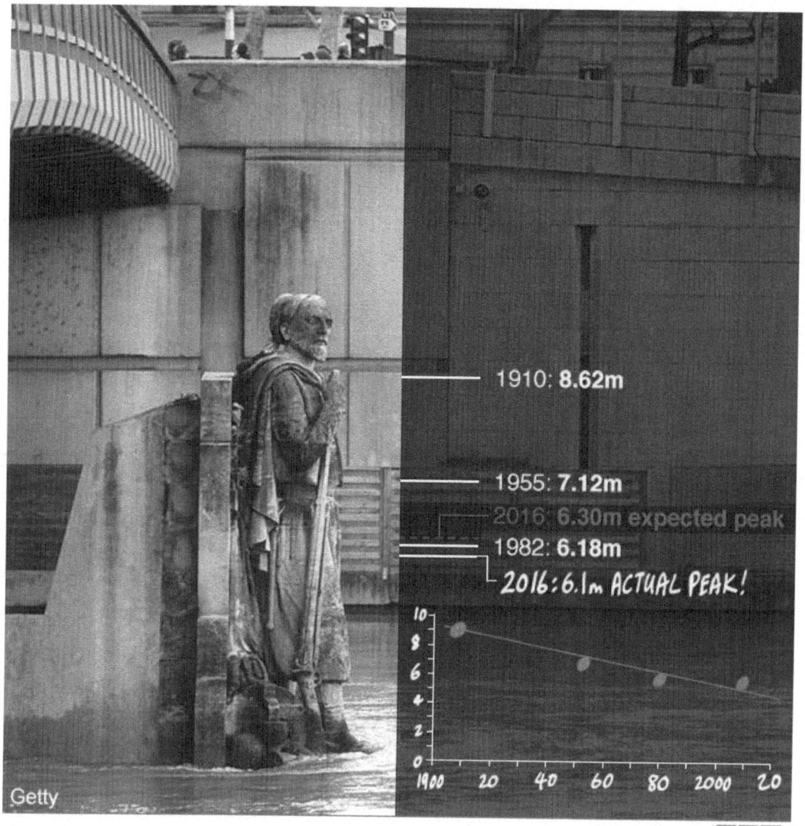


**[b] Paris Zouave statue as a flood guage**

**Figures 5.3[a & b]** Paris floods in Seine River – 1910 vs 2016

The so-called impacts on nature such as floods-droughts, cold waves heat waves, ice melt-glaciers retreat, sea level change [rise-fall], etc were there in the past and will be there in the future, as they are part of natural variability in climate change and also they are climate system-general circulation system specific activities. Natural variability plays the critical role in success and failure of Water resources availability and agriculture. This has lead to change in civilizations-migrations, changes in the course of mighty rivers, etc In the history over different parts of the globe. However, they are modified with natural disasters associated with volcanoes, earthquakes; etc coupled with modern human greed related activities over different parts of the globe.

Therefore, governments & UN agencies must encourage agroclimatological and agrometeorological studies using location-region specific data. There is an urgent need to consolidate the natural cycles in precipitation and temperature over different parts of the globe for better planning in water resources utilization and planning for agriculture.

4. **Truncated Data Series**: If a given data series present a cyclic variation, a truncated part of the cyclic variation present misleading inferences. This is exactly what is happening in many research reports using meteorological data series that were given hype by media world over. Indian Southwest Monsoon precipitation data and the global average temperature anomaly present 60-year cycle [presented in previous Parts of the book]. 60-year cycle means a sine curve of 60-years. In the sine curve by joining minimum to maximum it shows a steep rising trend; and by joining maximum to minim it shows a steep falling trend. The global average temperature anomaly presents both trend and superposed on it the 60-year cycle. The trend shows 0.6 °C per century rise. If we select the minimum to maximum period of sine curve, it shows there is steep increasing trend; the surface temperature data anomaly during the satellite period [1979 to date] shows exactly this. If we select maximum to minimum period of sine curve, it shows there is steep falling trend; exactly this has happened when a minister informed the Indian parliament that the southwest monsoon precipitation is decreasing.

5. **Pollution**: Individual governments and UN agencies must give top priority in reducing/controlling pollution [air, water, soil & food] and as well minimizing food waste. Pollution is the root cause of health hazards world over. In India adulterated food joins the pollution.

## References

King, J.W., 1975: Sun-weather relationship, Aeronautics & Astronautics, 13:10-19.

Parthasarathy, B. & Mooley, D.A., 1978: Some features of long homogeneous series of Indian summer monsoon rainfall, Mon. Wea. Rev., 106:771-781.

Parthasarathy, B., Munot, A.A. & Kuthawale, D.R., 1995: Monthly and seasonal rainfall series for al-India homogeneous regions and meteorological subdivisions 1871-1994, ISSN D252-1075, Research Report No. RR-065, IITM, Pune, India, pp.113.

Rao, K.N. & Reddy, S.J., 1972a: Lunar atmospheric tides in surface winds at six Indian stations, Indian J. Met. Geophys., 23:189-194.

Rao, K.N. & Reddy, S.J., 1972b: Solar and lunar atmospheric tides in rainfall at Poona, Indian J. Met. Geophys., 23:535-536.

Reddy, S.J., 1974: Lunar and solar atmospheric tides in surface winds and rainfall, Indian J. Met. Geophys., 25:499-502.

Reddy, S.J. & Rao, K.R., 1976: Radiation and evaporation distribution over India, The Natl. Geog.. J. of India, Vol. XXII, Parts 1 & 2, [I.G.U Special Number, Moscow], 54-63.

Reddy, S.J., 1977: Forecasting of onset of southwest monsoon over Kerala, Indian J. Meteorol Hydrol. Geophys., 28:113-114.

Reddy, S.J. & Rao, K.R., 1977: Effect of Solar flares on lower tropospheric temperature and pressure, Indian J. Radio & Space Phys., 6:44-50.

Reddy, S.J. & Lahori, S.N., 1977: Power spectral analysis of lower stratospheric meteorological data of H, T, u & v, Indian J. Radio Space Phys., 6:51-59.

Reddy, S.J., Juneja, O.A. & Lahori, S.N., 1977: Power spectral analysis of total & net radiation intensities, Indian J. Radio Space Phys., 6:60-66.

Reddy, S.J. & Rao, G.S.P., 1978: A method of forecasting the weather associated with western disturbances, Indian J. Met. Hydrol. Geophys., 29:515-520.

Reddy, S.J. & Singh, S., 1981: Climate and Soils of the semi-arid tropical regions of the world. In Proc. Summer Institute on Production Physiology of dry-land crops, APAU/ICAR, Rajendranagar, Hyderabad, India.

Reddy, S.J. & Neto, M.A., 1984: A method for the estimation of potential evapotranspiration and/or open pan evaporation over Brazil, Pesq. Agropec. Bras. (Brasilia),19:247-267.

Reddy, S.J., Neto, M.A. & Elpidio, M.G.S., 1984a: A simple method for the estimation of global solar radiation over northeast Brazil, Pesq. Agropec. Bras. (Brasilia), 19:391-405.

Reddy, S.J., Maite, R.K. & Seetharama, N. 1984b: An iterative regression approach for prediction of sorghum (sorghum bicolar) phenology in the semi-arid tropics, Agric. For. Meteorol., 32:323-338.

Reddy, S.J., 1984: Climate fluctuations and homogenization of northeast Brazil using precipitation data, pesq. Agropec. Bras. (Brasilia), 19:592-543.

Reddy, S.J., 1986: Climatic fluctuations in the precipitation data of Mozambique during the period of meteorological record, Comm. No. 39, Series Terra e Agua, INIA, Maputo, Mozambique, pp.40.

Reddy, S.J., 1987: The estimation of global solar radiation and evaporation through precipitation, Solar Energy, 38:97-104.

Reddy, S.J. & Mersha, E., 1990: Results: Climatic fluctuations in the precipitation data of Ethiopia during the period of meteorological record. Agric. Series 4, ETH/86/021 – WMO/UNDP, NMSA, Addis Ababa, Ethiopia, pp.28

Reddy. S.J., 1993: Agroclimatic/Agrometeorological Techniques: As applicable to dry-land agriculture in developing countries, www.scribd.com/Google Books, 205p [Book Review appeared in Agric. For. Meteorol., 67: 325-327, 1994].

Reddy, S.J., 1995a: Discussion: Over emphasis on energy in crop yield models, Agric. For. Meteorol., 77:113-120.

Reddy, S.J., 1995b: Comments: Sensitivity of some potential evapotranspiration estimation methods to climate change, Agric. For. Meteorol., 77:121-125.

Reddy, S.J., 2000: Andhra Pradesh Agriculture: Scenario of the last four decades, Hyderabad, India, 104p.

Reddy, S.J., 2002: Dry-land Agriculture of India: An agro-climatological and agrometeorological perspective, **BS Publications**, Hyderabad, India, 429p.

Reddy, S.J., 2007: Agriculture & Environment, Hyderabad, AP, India, 112p.

Reddy, S.J., 2008 & 2010, Climate Change: Myths & Realities & Annexure-I: Weather aberrations perspective of dry-land agriculture in Andhra Pradesh, www.scribd.com/Google Books, 176p & 114p.

Reddy, S.J., 2011: "Green" Green Revolution: Agriculture in the perspective of climate change, www.scribd.com/Google Books, 160p.

Reddy, S.J., 2014: Water logging and water productivity in agriculture, Proc. 4th International Conference on "Hydrology and watershed management" [ICHWAM-2014], Allied publishers Pvt. Ltd., pp.683-692 [Vol. II, first article].

Reddy, S.J., 2016: Irrigation and Irrigation Projects in India: Tribunals, Disputes and Water Wars Perspective, **BS Publications**, Hyderabad, India, 132p.

WMO [World Meteorological Organization], 1966: climate change, Geneva, Switzerland, WMO Tech. Note No. 79, WMO, 195 TP 100, pp81.

blog sites: my comments on --

DotEarth/The New York Times/Environment

The Guardian

watts up with that

# Index

## S



## T



## U



## V

## W



## Z

9 789352 301515

Printed at Repro India Ltd.